Information Systems Management

Edited by Rohit Raja and Hiral Raja

Published in London, United Kingdom

Information Systems Management
http://dx.doi.org/10.5772/intechopen.100784
Edited by Rohit Raja and Hiral Raja

Assistants to the Editors: Parul Dubey and Deepika Ingle

Contributors
Gopinath V., Aondover Eric Msughter, Aondover Ogechi Perpetua, Adebija Lillian Itiafa, Roman Rak, Dagmar Kopencova, Anil Vaidya, Mihai Caramihai, Cristina-Maria Vipie, Alexandra-Diana Afumatu, Aouatef Safo Mahani, Bapan Kumar Maity, Sutapa Paul, Massimo Bianchi, Rahul K Patel, Ankur Solanki, Mansi Patel, Jeny Gosai, Nitin Chaudhari, Fernando Cassinda Quissanga

First published in London, United Kingdom, 2023 by IntechOpen
IntechOpen is the global imprint of INTECHOPEN LIMITED, registered in England and Wales, registration number: 11086078, 5 Princes Gate Court, London, SW7 2QJ, United Kingdom

British Library Cataloguing-in-Publication Data
A catalogue record for this book is available from the British Library

Additional hard and PDF copies can be obtained from orders@intechopen.com

Information Systems Management
Edited by Rohit Raja and Hiral Raja
p. cm.
Print ISBN 978-1-80355-651-2
Online ISBN 978-1-80355-652-9
eBook (PDF) ISBN 978-1-80355-653-6

For EU product safety concerns:
IN TECH d.o.o., Prolaz Marije Krucifikse Kozulić 3, 51000 Rijeka, Croatia,
info@intechopen.com or visit our website at intechopen.com.

Meet the editors

Dr. Rohit Raja received his Ph.D. in Computer Science and Engineering from Dr. CV Raman University, India, in 2016. His main research interests include face recognition and identification, digital image processing, signal processing, and networking. Presently, he is an associate professor in the IT Department, Guru Ghasidas Vishwavidyalaya (A Central University), India. He has authored several journal and conference papers. He has good academic and research experience in various areas of Computer Science and Engineering (CSE) and information technology (IT). He has filed and successfully published twenty-nine patents. He has received many invitations to be a guest at the Institute of Electrical and Electronics Engineers (IEEE) conferences. He has published 100 research papers in various international and national journals and proceedings at reputed international and national conferences. He has been nominated to the board of editors and reviewers of many peer-reviewed and refereed journals.

Dr. Hiral Raja is an associate professor in the Mathematics Department at Dr. C.V. Raman University, India, where she received her Ph.D. in Mathematics in 2017. Her main research interests include image processing, embedded system, artificial intelligence, and sensor signalling. She was successfully granted two national and two international patents. She has published approximately ten research papers in national and international journals and edited one book.

Contents

Preface **IX**

Section 1
Introduction **1**

Chapter 1 **3**
Software Development Management
by Rahul Patel

Chapter 2 **17**
Classification in Multi-Label Datasets
by Aouatef Mahani

Chapter 3 **31**
Research Trends in Library and Information Science in India during 2011 to 2018
by Sutapa Paul and Bapan Kumar Maity

Chapter 4 **53**
Network Analysis in the Information Systems Management: Implications for a Transdisciplinary Approach
by Massimo Bianchi

Section 2
Information Systems and Their Applications **67**

Chapter 5 **69**
Secured Storage Mechanism for the Cloud-Based Banking Applications Using ECC
by Gopinath V.

Chapter 6 **87**
Impact of Digital Vehicle Identification Errors on Critical Information Systems
by Roman Rak and Dagmar Kopencova

Chapter 7 107
Blockchain-Based Educational Certificates: A Proposal
by Cristina-Maria Vipie, Alexandra-Diana Afumatu and Mihai Caramihai

Chapter 8 129
Artificial Intelligence and the Media: Revisiting Digital Dichotomy Theory
by Aondover Eric Msughter, Aondover Ogechi Perpetua and Adebija Lillian Itiafa

Chapter 9 143
Hybrid Perovskite-Based Memristor Devices
by Mansi Patel, Jeny Gosai, Nitin Chaudhari and Ankur Solanki

Chapter 10 159
Technology Strategy Formulation for Global Corporations
by Anil Vaidya

Chapter 11 175
Comparative Study of Information Security in Mobile Operating Systems: Android and Apple iOS
by Fernando Cassinda Quissanga

Preface

One of the most important tools accessible to company managers for attaining operational excellence, generating new goods and services, enhancing decision-making, and gaining a competitive edge in information systems. This book provides a current and complete introduction to information systems utilized by businesses today. It is for the technical reader who deals with huge amounts of data.

The book includes chapters on software development, cloud implementation, networking, managing massive datasets, and more. Technologies like blockchain and artificial intelligence (AI) are the backbone of automated systems. Using these foundational areas of IT, the authors provide their perspectives on information management.

Chapter 1, "Software Development Management", discusses how to choose the correct model for software development, ensuring that its principles will be followed and enforced at every stage of the process. Organizations may benefit from software development management (SDM) because it ensures their applications are being developed and maintained in a logical fashion and that they are producing software solutions that satisfy the needs of all their stakeholders, both internal and external.

Chapter 2, "Classification in Multi-Label Datasets", discusses classification, which is the most often performed activity in data mining. It focuses on multi-label classification, presenting problem transformation strategies and algorithm adaptations.

Chapter 3, "Research Trends in Library and Information Science in India during 2011 to 2018", is a bibliometric analysis of PhD theses submitted to the Sodhganaga repository between 2011 and 2018. The results show that academic attention is shifting from fundamental library and information science (LIS) subjects to IT and computer applications.

Chapter 4, "Network Analysis in the Information Systems Management: Implications for a Transdisciplinary Approach", looks at how information flows in the digital world by analyzing the processes involved in this flow using information-measuring tools.

Chapter 5, "Secured Storage Mechanism for the Cloud-Based Banking Applications Using ECC", discusses elliptical curve cryptography (ECC) as a safe storage mechanism for cloud-based financial apps.

Chapter 6, "Impact of Digital Vehicle Identification Errors on Critical Information Systems", examines the role a high-quality Vehicle Identification Number (VIN) plays in ensuring a vehicle will be used safely. Accident victims' safety may be jeopardized if emergency responders waste time trying to figure out who the car's owner is due to incorrect identification, which may slow down or compromise the quality of their work.

Chapter 7, "Blockchain-Based Educational Certificates: A Proposal", discusses the use of blockchain databases to store academic transcripts. Most such systems today use blockchain as a safe means of verifying and exchanging student information, academic certificates, and institution databases.

Chapter 8, "Artificial Intelligence and the Media: Revisiting Digital Dichotomy Theory", reviews the impact of AI on the media industry in which it is revolutionizing various aspects of content creation, distribution, and consumption.

Chapter 9, "Hybrid Perovskite-Based Memristor Devices", discusses the development of modern electronic devices for advanced applications driven by AI and machine learning (ML). It highlights the increasing demand for "universal memory" devices with exceptional qualities. The chapter explores the history and current state of memristor technology, including novel structures, switching mechanisms, materials, and various applications such as data storage, artificial synapse, light-induced resistive switching, logic gates, and mimicking human behavior.

Chapter 10, "Technology Strategy Formulation for Global Corporations", discusses technology strategy formulation for global corporations, considering the social and political factors that arise due to cultural and geographical diversity across different countries and locations. The author emphasizes that focusing solely on technology and economics is not sufficient for these corporations, as the execution of technology strategies is influenced by social structures and political dynamics.

Finally, Chapter 11, "Comparative Study of Information Security in Mobile Operating Systems: Android and Apple iOS", is a comparative study of Android and iOS mobile operating systems that finds through qualitative and quantitative analysis that iOS offers stronger information security compared to the Android system. Although virus infections are less prevalent in iOS, risks exist from other cyber-attacks like spyware capturing sensitive information. Cybercriminals target both platforms, particularly social networks. The study recommends security measures such as strong passwords, advanced encryption, firewall and privacy tools, document encryption, and encrypted communication apps like Signal to enhance overall security.

Dr. Rohit Raja
Associate Professor,
IT Department,
Guru Ghasidas Vishwavidyalaya (A Central University),
Bilaspur (CG), India

Dr. Hiral Raja
Associate Professor,
Mathematics Department,
Dr. C.V. Raman University,
Bilaspur (CG), India

Section 1

Introduction

Chapter 1

Software Development Management

Rahul Patel

Abstract

The software development and management (SDM) practice helps organizations to ensure that their software products are developed methodically and delivered in accordance with the stakeholders' requirements. The main purpose of the SDM is to map out the management tasks and sequences them rationally. Software must be developed in a systematic manner to ensure that it is developed on time, within budget and with all functional and non-functional capabilities. SDM practice plays a significant role in ensuring that the software is developed in a systematic manner, traditionally using the models such as waterfall model, agile model, and V-model. There are many modern software development models such as DevSecOps that extends the capability of existing models and makes it more cloud friendly. Web-based development and delivery of software applications as well as Low-Code/No Code (LCNC) are becoming increasingly popular and valuable due to availability of tools and technology for SDM. Selecting the right model for developing software ensures that all the disciplines of that model are implemented and monitored during the SDM process. Hence, SDM practices strategically manage how software is being developed, tested, and deployed and creates the business value for the stakeholders.

Keywords: software development management (SDM), cloud development, software development models, DevOps, DevSecOps, continuous practices, No-Code/Low-Code development

1. Introduction

The term 'software' has been used for many kinds of computer programs such as a single program or suite of programs ranging from operating systems, middleware, environment, database, desktop, mobile, embedded or web applications. All software products have many unique perspectives such as functionality, development environment, development environment, platforms, users, and way it is being developed and used and hence any single software development and management framework cannot be effectively used for all kinds of software products. Software development and management is a process of developing and managing the required software by adopting various well defined software development life cycle stages. The SDM practice must emphasis on making software reliable, high-quality, and easy to use for users [1]. There are many software-development models that can be used to manage the software development process for achieving these goals however methodology should be selected carefully. Unique challenge presented is that selected SDM must support modern computing environment such as web and cloud where various sub operating environment are dynamic and not fully controlled by any one organization. Deciding which model to use for developing a certain type of software is a complex task. The software developers and managers need to work closely to acquire some knowledge about the software project and then decide which model would work best for the team. The success and failure of the project depends on the selection of the software model [1, 2]. Hence, selecting a proper software model and managing all associated processes and artifacts through various phases of development, are the keys for the effective software development and management. These models articulate the various stages of the process and the order in which the process must be carried out to achieve the required objectives. This paper discusses the various technical and management activities, workflows of the models, and how the models can be improved for future use. Modern and innovative approaches such as continuous integration and Continuous development, and extreme programming are discussed in detail.

2. Purpose

The main purpose of the Software Development and Management Practice is to provide an effective and efficient method for managing software development cycle. SDM practice should be able to manage and address all functional, technical, warranty, operational, and environmental requirements. In addition, the SDM practice must be able to deliver software products efficiently on time and within budget. These goals can only be achieved by applying a systematic approach to manage the software development lifecycle. Software development and management provide an effective framework to develop and manage software applications. The software development and management process will help in effectively planning and managing the process before starting the actual development process [3]. Such methodical approach is efficient and effective in reducing unnecessary costs during the development process which enables the team to estimate the cost of the project. Since such approach allows a systematic process of testing the software before the software is rolled out, the developers can design and build high-quality software products. Hence, the purpose of software development and management is to enhance the software product by evaluating the effectiveness of the software. In modern application development and management practices we must consider effectiveness and efficiency of the approach when the use of such applications software is cloud or web based as well as traditional uses cases such as internally hosted applications.

3. Scope

Software development and management practice covers many different aspects such as design, architecture, development, testing, and finally deployment. The scope of software development and management includes the following architecting the software solution, create a solution design blueprint, software development, testing the solution from various perspective such as unit testing, integration testing, regression testing, software security testing, and user acceptance testing. Stabilization also calls for managing the software product with code repositories, packaging, version control, and patching are also part of the SDM practices. Unique considerations are that such software are deployed and managed thru the cloud and web where many distributed components are not under the control of the organizations.

Scope: After the initiation of the project between the client and the organization, the project manager of the organization assigns the project team. Along with the clients, the organization defines the project scope by using requirement gathering and analysis.

Development: In this stage, the developers set up the development environment framework and develops the code based on requirements decided during the requirement gathering and analysis phase of scoping activity. Developer in this stage also ensures that the software is fit for the purpose.

Stabilization: In this stage, developers focus more on the quality of the software and obtains independent feedback from the testing team. Testing teams carry out various functional and non-functional tests to determine the quality of the software and ensures that the software is fit for the use.

Delivery: The final version of the software is delivered to the customer after the software is tested and the bugs are fixed. Knowledge transfer occurred during this phase with initial support, training, and documentation related with the software solution.

Maintenance and Support: After the software is accepted and released, the ongoing support for troubleshooting, escalated incident support, bug/fix, work arounds, and patching are provided by the SDM team until the end of life for the software solution.

4. Workflow for software development management

The Software development and management process follows the software development life cycle (SDLC) to develop the software within the organization. The SDLC has a detailed plan of how to develop, maintain and upgrade the overall development process. Workflow depicted in the **Figure 1** seems sequential, however, it is hardly that simple. Most management approaches include all of the steps but in many different spiral, iterative and analogous combination since such chronological workflow is not effective and efficient in faces changing environments. However tasks carried out in each phases are still relevant and effective.

4.1 Planning and requirement analysis

In this stage, the project managers obtain the requirements related inputs from the clients, sponsors, and different stakeholders of the organization. The data gathered from various stakeholders are used analyze the feasibility of the product in affordable, functional, and specialized areas and to develop the design and the approach that would work best for the scope and within the constraints [4]. These inputs are

Figure 1.
Workflow of software development and management.

used for developing the project detailed plans and milestones. Consistent process of gathering requirements at depths and breadth is essential to succeed in software projects since they go through many levels of translations from business requirements to technical requirements.

4.2 Defining requirements

The main reasons for variation in the estimation for time and budget in the software projects are related to the lack of complete understanding of the requirements properly. These gaps must be minimized at the time of requirements analysis so that the variations in required time and money to complete the project can be minimized. Source of customer dissatisfaction is also related to the lack of understanding the need. Specific requirements are defined and documented at the end of the analysis of the requirements. Requirements are categorized as functional, operational, interoperability and technical environment related requirements. The clients then look at the requirements and approve the requirements. The finalized document is called the Software Requirement Specification (SRS) which consists of all the software requirements that must be documented, planned, and tracked throughout the life cycle of the project.

4.3 Designing

Many technical design guidelines and standards are decided or reused from previous projects based on common software engineering principles. Fundamental software engineering concepts, such as abstraction, modularity, reusability, compatibility, extensibility, scalability, and maintainability are important aspects that are considered in design phase to develop the high quality and longer-lasting software solutions [2]. Level of modularity needed in the software is decided in the initial phase of the design to allow the level of flexibility for future changes. Furthermore, opportunities for code reusability, write once use many times, are also considered in the designing phase to increase efficiency and quality of the software product. Nonfunctional requirements such as compatibility, scalability, and maintainability are considered to make software more adoptable, change friendly.

In the designing phase of the workflow. The software architects plot the best architecture for the software to be developed using the SRS, that was developed in the previous step. To give a better context to the requirements, user stories or use cases are developed that would clarify interfacing requirements such as inter-functional inputs and output requirements. Software architects usually design one or more architecture and document it in the Design Document Specification (DDS). This document is further scrutinized with all the stakeholders with respect to the project boundaries, hazard evaluation, time imperatives and the financial plan.

4.4 Development

The actual development starts, and the software is built in this stage based on the DDS defined in the previous step the programming code for the software solution is generated. Various requirements and related user stories are itemized and coordinated within distributed and parallel development groups. With such approach programming and application development process, unknowns are minimized, and development become more efficient and effective. During the generation of the code, the developers follow organization wide coding standards and programming tools needed for the programming such as the interpreters, compilers, and debuggers [5]. Even though there are separate dedicated measures for testing and quality assurance, unit testing is performed at this stage to detect and address defects at the unit level. Unit testing is performed to ensure that an application or a product meets its design and behaves as per the requirements. These tests are run by the developers in their development environment. The developers test the internal logical structure of each software component as well as the individual function or procedure.

4.5 Testing and quality assurance

Software Testing is performed to detect the defects in the product while quality assurance process addresses the need for preventive measures to avoid the defects in the product. Quality Assurance measures ensure that the developed solution addresses all the requirements as defined in agreed specifications, standards, and functionality without known bugs, defects, or abnormal behaviors [6].

Software solution is tested from various perspectives before deployment for detecting defects and bugs. The defects, bugs, and abnormal behaviors in the software are discovered, tracked, fixed, and retested until the software reaches the quality requirements, that are characterized in the SRS.

Various types of tests conducted during this phase are integration testing, regression testing, software security testing, and user acceptance testing. Functional testing focuses on the functional requirements and software behavior in accordance with pre-determined requirements. Full set of functional inputs are used to validate the set of output it generates to affirm that all application functionalities work appropriately as indicated by the business requirements. Regression testing is performed to make sure previously developed, and tested application performs the same after making changes or enhancement in the application by retesting all functional and non-functional test cases. Integration testing is the part of programming testing where individual programming modules are amalgamated and tested overall to verify the coupling and interactions between them. Interface testing is conducted to make sure all human to software and all software-to-software interfaces are as per the requirements. Interface testing includes the UI testing, API testing, web services testing. Finally, the user acceptance testing is the last phase of testing in the software development life cycle where the business users or functional experts performs the acceptance testing to make sure the application meets all the customer requirements.

4.6 Deployment

Once the product is tested and ready to be conveyed, it is distributed and delivered to the end users. In this stage, various components of the software are packaged in a way that deployment checks the perquisites and then the components are sequenced, deployed, and verified. Various deployment mechanisms are decided and operationalized in this phase such as Manual (software is installed manually on the end points) v/s automated (software is installed automatically from centralized deployment systems), Push (software is pushed to targeted endpoints) v/s Pull (targeted endpoints may pull the software based on their preferences), Phased (deployment occurs in logically divided phases) v/s Big-Bang (app endpoints receive software at once). Subsequently, the maintenance, updates, and bug fixes are deployed using similar methods of deployment.

5. Workflow for software development management

There are many Software Development life cycle (SDLC) models that articulates series of steps, tasks, and dependencies for developing software [7]. Effectiveness of the model used for software development depends on many aspects such as requirements from users and developers, clarity and flexibility of scope changes, team size and dynamics, and need for interactions between stakeholders [8]. It is very important to consider time, cost, quality, product, risk, and resources while selecting and implementing software development model [1]. Software development models are very flexible and hence it is not necessary to stay within the bounds of each model. Various well recognized and followed models that can be modified and applied for the best results are as follows:

5.1 Waterfall model

The waterfall model is the oldest model for software development. The result of one stage can be inserted to the next stage successively in this model. Each phase is started only when the goals of the previous stage are achieved. Hence, the stages do not intersect in the waterfall model. The stages in this model are requirement analysis, System design, Implementation, Integration and Testing, Deployment of System and Maintenance. The waterfall model is straightforward. The stages are clearly categorized, and the processes and requirements are well archived. However, this model is not practical for modern complex and dynamic software development projects due to its lack of flexibility for accommodating changing requirements.

5.2 V-model

The V-model is called the verification and validation model where the execution of cycles occurs in a successive way in V-shape [1]. This model is an expansion of the waterfall model in which each testing stage is related with the subsequent development stage. The upside of V-model is that it is straightforward. However, this model is not that flexible to change as the changes are common in the dynamic world. In this model, the stages are completed in turns with verification and validation for that specific stage. One of the biggest benefits of this model is that defects are found and resolved at the early stages, which saves time later. This model functions well on modest activities where requirements do not change and are surely known. Each stage has a particular output to audit the process.

5.3 Iterative model

In this model, the implementation of a small set of requirements is done initially. In turn, iteratively, all the remaining requirements are implemented until the complete system is implemented and the software is ready to be deployed. Here, advancement starts by executing a piece of the software product that is needed to recognize the future necessities. A similar cycle is rehashed to create another form of the product toward the end goal. At every iterative cycle, the changes for the plan are developed and new functionalities are delivered. Henceforth, the basic idea of the iterative model is to foster the product through iterative cycles and incremental portions all at once. The iterative model is well suited when the requirements have interdependencies and they become clear with time and intermediate results. It is also well suited for large or long projects where requirements can be prioritized, and system can be functional with only partial requirements and products can be delivered in iteration effectively reducing the time to market.

5.4 Spiral model

The spiral model is an amalgamation of the ideas of an iterative model and waterfall model. In this model, the team addresses the requirements with incremental releases of the software through iterations around the spiral effectives decreases pending requires to be addresses in future pending spiral iterations. Instead of predefined set of requirements for each iteration and development stages, the spiral model is designed to empower development teams to select the most appropriate approach given the current goals, constraints, and risks [9]. Furthermore, the number of spiral iterations and requirements addressed in each iteration are not fixed and hence, the model allows additional flexibility to the development teams for take the most appropriate course of actions based on situation. Decreasing number of pending requirements with each iteration creates a spiral effect. Spiral model-based development can be effective in a variety of situations such as varying level of risks, project length, scope changes, as well as financial resources.

5.5 Agile model

This model is a mix of iterative and spiral process which centers around the versatility of the software and the satisfaction of the customers so that the software can be delivered at the quicker rate. Agile models are people-centric, recognizing the value competent people, and their relationships bring to software development [10]. In addition, it focuses on providing high customer satisfaction through efficient delivery of high-quality software, active participation of stakeholders, and continuous change [11]. This model allows breaking the software into small incremental modules for each iteration. Each iteration of the product lasts for the short term, typically less than a month, where development teams work with stakeholders on aspects such as planning, requirement analysis, design, coding, and testing simultaneously [10]. The working software is tested and demonstrated to the clients and stakeholders as a final step of each iteration. The agile model helps development teams divide the software into small modules that can be built within small timeframe and delivers specific high-quality deliverables for every release cycle. Agile model is one of the most effective methodologies for collaborative organizations.

5.6 RAD model

The Rapid Application Development model requires no preparation and development depends on prototyping and iterative turn of events. This model is very collaborative in nature where the information related to requirements and use cases are gathered from the customers through workshops and focus groups using the iterative concepts or reusing the existing prototypes. Here, the functional models are developed as prototypes and further development efforts are coordinated leading to the final product. Since no pre-planning processes are required, it is easy to include the changes in the software development process in iterations. The advantage of using this model is that the prototypes can be delivered, reused, and improved in stages. RAD model is aimed at speeding software development process. There are many RAD based implementations such as extreme development (XP) where business champion and developers lead efforts in increments, Join Application Development (JAD) where customers, end users, and developer work together in workshops, Lean Development (LD) where requirements are kept lean to include only minimum or must-have functionalities, and Scrum where software is developed in multiple short iterations.

6. Roles and responsibilities

Models discussed in previous section are not the complete list and they are not used exactly as they are described however the practical approach may match most of the framework of a specific model. There is proliferation of SDLC models due to many unique dependencies and varying demands of applications. However, to be successful regardless of the model used, clear role and responsibilities need to be defined. These roles and responsibilities are assigned to different groups of people based on which model is being used for software development. Common roles and responsibilities of people involved in Software Development and Management process is outlined in this section (**Table 1**) [12]:

Roles	Responsibilities
Project Sponsor	• Provide governance framework for the project team • Provide Budget and Requirements • Approve project charter and overall timeline • Review project progress periodically • Direct management team if/as needed
Project Manager	• Supervise the activities of all the project teams • Manage risks and risk mitigation efforts • Manage tactical changes in scopes and schedule as/if needed • Manage the financial resources and timeline • Allocate tasks and resources • Provides status updates to stakeholders and customers. • Develops the software project plan • Ensured that delivery on time and within budget • Determine phases and milestones of the project • Implement effective project management methodology
Subject Matter Expert	• Provide technical leadership with expertise in SDM • Lead with knowledge of the discipline and technology • Translate the business requirement into technical requirements • Provide recommendations for improvements. • Decide and implement most effective SDLC model

Roles	Responsibilities
Product Owner	• Represent the business and end-users • Work closely with the user group to understand the requirements needed for the release. • Prioritize the backlog during the development • Make requirements decision to maximize the return on investment • Make user acceptance and production release decisions
Technical Lead	• Translates the business requirements to technical solutions. • Lead the software developers and propose the solution. • Communicate the status of the development efforts and challenges • Establish discipline within the software development team. • Provides solution and support for escalated challenges and technology risks
Software Developer	• Develop the solution addressing the technical requirements • Resolve or escalate development challenges and risks • Integrates the software components by developing effective and efficient software programs • Resolve bugs and errors or provides work arounds as needed
Software Tester	• Test and ensure that the results meet the business requirements, and the results are free of bugs and errors • Create the test plans, designs, test cases, and test data • Make the success/failure decision based on test specifications, defect reports, and the test results • Execute all types of tests such as system testing, regression testing and integration testing

Table 1.
Roles and responsibilities for software development and management.

7. Newer approaches for software development and management

The Software Development and Management process can be enhanced to be more effective using advanced methods and technologies. Automations, collaborations, Security, Cloud, and Mobile platforms are few areas that are driving innovative ways of software development and management that are more effective [9, 13]. Some of the modern approaches are listed below:

7.1 Development and operations (DevOps) approach

DevOps means development and operations as a single unified approach for SDM with effective communication between development and operational team. DevOps process creates a work environment with development, testing, and operations are part of a continuous cycle. The DevOps process comprises - planning, coding, building, testing, releasing, deploying, using, and providing the feedback with monitoring by restarting the loop. In DevOps approach, the IT team writes the software that meets the user requirements, deploys, and runs optimally. DevOps approach extends the software development management cycle to include continuous monitoring and continuous improvement.

DevOps ideas is now extended to include security to make it DevSecOps where the idea is to integrate security as part of development and operations workflows. DevSecOps mandates that security should be a key component of the application development pipeline since risk of security incidents related to the code vulnerabilities are becoming very costly [14]. Security is integrated in each step of the workflow. DevSecOps include security by providing the Secured Development Environment, including secure coding practice, static and dynamic security testing, and continuous monitoring the operational environment to identify and reporting security challenges.

7.2 Optimizing the workflow

The development teams usually follow the workflow that was established by the reference development model. However, it is uncertain that the previously selected workflow model works efficiently to develop the current software. Selection of the model must be analyzed periodically to identify slow stages and repetitive tasks, error prone processes, and variation in the quality of the software. Such indicators help to identify and resolve the internal management issues and can help in optimizing the workflow for the software development and management processes.

7.3 Improving the code quality and accuracy

The code quality and accuracy can be improved by allocating time for testing the software from the beginning. These goals can be achieved by using automated test cases and monitoring small fragments of the solution and making sure that we are progressing to develop the fundamentally sound software products. Even though, implanting such testing steps, widen the software development process since such tests are performed automatically, it improves quality of software products without slowing down the overall process. Many functional, non-functional, security, and technology related tests can be included to improve accuracy and quality of the code while being developed.

7.4 Cloud based Software development

The cloud-based information technology services have become a new platform for developing modern software applications that can run on advance platforms such as serverless computing. An early movement toward using web environments for application development came from companies that delivered software products as a service [15]. Software development and management decisions are made early not only to implement development and management methodologies but selection of development environment as cloud friendly programming languages, serverless coding, continuous integration, and continuous delivery mechanisms.

Another developing area of software development is Low-Code/No-Code (LCNC) software development. LCNC approach is based on automated no-code tools allowing used use develop application based on configuration driven, APIs based, cross-platform in cloud-based application development [16]. Many integration points and custom requirements still require low code development however underlying assumption is that many software building blocks are readily available from the cloud vendors or opensource that enables software development to be No Code or Low code effort. NCLC is still evolving field and hence the LCNC based solutions are very generic and tend to create bulky code that lacks runtime efficiency [C]. As a result, NCLC based software development cycles still needs code optimization and review to make it more efficient and through testing.

7.5 Continuous practices – integration (CI) and delivery (CD)

Modern application development and management practices need to devote significant resources to developing and delivering high-quality software at a faster rate [2]. Continuous Integration (CI) and Continuous Delivery (CD), referred to as continuous practices, are modern SDM practices aimed at assisting organizations in accelerating the development and delivery of software features without sacrificing quality. While CI encourages integrating work-in-progress numerous times a day, CD is concerned with the capacity to release values to consumers rapidly and reliably by including as much automation as feasible. CD includes the deployment to the live environment without human intervention.

Continuous integration is a software development practice where multiple members of a development team work on a various branch of the software simultaneously and use a version control system to frequently integrate their work to the main branch [17]. Each change is built and verified to detect integration errors as quickly as possible. Continuous delivery is a software development methodology where the release process is automated. Every software change is automatically built, tested, and deployed to production. Before the final deployment and delivery, a person, an automated predefined test, or a business rule decides when the final push should occur [17]. Security weaknesses and vulnerabilities can pass undetected through oversights and mistakes in programming and testing, making code susceptible to vindictive activity. Thus, security best practices we discussed so far should be a part of the CI/CD pipelines by using weakness checkers and as separate additional security checks during the testing system. CI/CD required test automation tolls and deployment automation tools. Test automation tools are used for continuously unit, functional, and performance testing when any change is introduced in the code to ensure the stability of the code base [18]. Similarly, for Delivery and deployment, automation tools are used for packaging and integrating code changes in the live environment at much faster rate without any human intervention and by adhering to all preprogrammed and automated business rules [18].

Continuous practices have few benefits such as receiving more and faster feedback from the software development process and customers; [19] having frequent and reliable releases, which leads to improved customer satisfaction and product quality; and [17] strengthening the connection between development and operations teams and eliminating manual tasks [20, 21]. Continuous practices are making inroads in software development industrial practices across diverse domains and sizes of enterprises. Adopting continuous practices, on the other hand, is not a simple undertaking since organizational processes, practices, and tools may not be prepared to support the very complex and hard nature of these practices [22].

8. Conclusion

Ensuring that high quality systems are delivered, assisting the projects by using strong management controls, and maximizing the productivity of the team members are the primary objectives of software development and management. To meet these objectives, the requirements, which are needed are: supporting the project teams of the various scopes and types, assisting the technical and the management teams, providing guidance of how to install the software which is needed for the development of the software. The technical activities include analyzing, designing, coding, testing, installing the software's and defining the technical strategy of the project. The management activities include defining the project objectives, tracking the progress of the project, reporting the project status, risk assessment, quality assurance and user interaction post implementing the software. Hence, to meet all these objectives and requirements many software models are used. These models have a layered approach to analysis, design, installation, and support. It also explains how the software must be developed by creating tasks, getting the right deliverables, and organizing the resources where they can be quickly accessed. Defining and selecting the software models should be taken up and managed with appropriate levels of expertise. So, the software development and management constitute the culture of the software teams and how the development process is planned, managed, and executed in a professional manner.

Author details

Rahul Patel
Illinois Institute of Technology, Chicago, United States of America

*Address all correspondence to: rpatel37@iit.edu

References

[1] Alam I, Sarwar N, Noreen I. Statistical analysis of software development models by six-pointed star framework. PLoS One. 2022;**17**(4):1-17. DOI: 10.1371/journal.pone.0264420

[2] Adnan NH, Ritzhaupt AD. Software Engineering Design Principles Applied to Instructional Design: What can we Learn from our Sister Discipline? TechTrends: Linking Research & Practice to Improve. Learning. 2018;**62**(1):77-94. DOI: 10.1007/s11528-017-0238-5

[3] Ping L. A quantitative approach to software maintainability prediction. In: 2010 International Forum on Information Technology and Applications. Kunming, China: IEEE; 2010. DOI: 10.1109/ifita.2010.294

[4] Moreno W, Afonso P, Costa H. Challenges and solutions of Project Management in Distributed Software Development. In: 2019 XLV Latin American Computing Conference (CLEI). Panama, Panama: IEEE; 2019. DOI: 10.1109/clei47609.2019.9089043

[5] Hegedus P, editor. Revealing the effect of coding practices on software maintainability. IEEE International Conference on Software Maintenance. 2013;**1**(1):578-581. DOI: 10.1109/icsm.2013.99

[6] Iacob IM, Constantinescu R. Testing: First Step Towards Software Quality. Journal of Applied Quantitative Methods. 2008;**3**(3):241-253

[7] Sharma P, Singh D. Comparative study of various SDLC models on different parameters. International Journal of Engineering Research. 2015;**4**(4):188-191. DOI: 10.17950/ijer/v4s4/405

[8] Ben-Zahia MA, Jaluta I. Criteria for selecting software development models. 2014 Global Summit on Computer & Information Technology (GSCIT). Sousse, Tunisia; 2014. DOI: 10.1109/gscit.2014.6970099

[9] Hendrix TD, Schneider MP. NASA's TReK Project: A Case Study in Using the Spiral Model of Software Development. Communications of the ACM. 2002;**45**(4):152-159. DOI: 10.1145/505248.506004

[10] Nerur S, Balijepally V. Theoretical Reflections on Agile Development Methodologies. New York, NY, United States: Communications of the ACM. 2007;**50**(3):79-83. DOI: 10.1145/1226736.1226739

[11] Siau K, Erickson J. Advanced principles for improving database design, Systems Modeling and software development. NY, United States: Information Science Reference; 2009. ISBN-13: 978-1605661728

[12] Yilmaz M, O'Connor RV, Clarke P. Software development roles. ACM SIGSOFT Software Engineering Notes. 2015;**40**(1):1-5. DOI: 10.1145/2693208.2693239

[13] Velmourougan S, Dhavachelvan P, Baskaran R, Ravikumar B. Software development life cycle model to improve maintainability of software applications. In: Fourth International Conference on Advances in Computing and Communications. Cochin, India: IEEE; 2014. DOI: 10.1109/icacc.2014.71

[14] Scanlon T, Morales J. Revelations from an agile and DevSecOps transformation in a large organization: An experiential case study. Proceedings of the International Conference on

Software and System Processes and International Conference on Global Software Engineering. 2022;**1**(1):77-81. DOI: 10.1145/3529320.3529329

[15] Cusumano MA. Technology Strategy and Management: The Cloud as an Innovation Platform for Software Development: How cloud computing became a platform. Communications of the ACM. 2019;**62**(10):20-22. DOI: 10.1145/3357222

[16] ElBatanony A, Succi G. Towards the no-code era: A Vision and plan for the future of software development. In: Proceedings of the 1st ACM SIGPLAN International Workshop on Beyond Code. Chicago, Illinois, United States: ACM; 2021;**1**(1):29-35. DOI: 10.1145/3486949.3486965

[17] AWS. Continuous delivery and continuous integration - AWS codepipeline. n.d. Available from: https://docs.aws.amazon.com/codepipeline/latest/userguide/concepts-continuous-delivery-integration.html [Accessed: October 17, 2022]

[18] Donca I-C, Stan OP, Misaros M, Gota D, Miclea L. Method for Continuous Integration and Deployment Using a Pipeline Generator for Agile Software Projects. Sensors. 2022;**22**(12):4637. DOI: 10.3390/s22124637

[19] Humble J, Farley D. Continuous delivery: Reliable software releases through build, test, and Deployment Automation. Upper Saddle River, NJ, United States: Addison-Wesley; 2015. ISBN: 978-0321601919

[20] Laukkanen E, Itkonen J, Lassenius C. Problems, causes and solutions when adopting continuous delivery—a systematic literature review. Information and Software Technology. 2017;**82**:55-79. DOI: 10.1016/j.infsof.2016.10.001

[21] Rodríguez P. Continuous deployment of software intensive products and services: A systematic mapping study. Journal of Systems and Software. 2017;**123**:263-291. DOI: 10.1016/j.jss.2015.12.015

[22] Chavan N, Bharambe N, Deshmukh S, Ahire D, Jain AR. Implementing DEVSECOPS pipeline for an enterprise organization. International Journal of Advanced Research in Science, Communication and Technology. 2020;**3**:46-72. DOI: 10.48175/ijarsct-4883

Chapter 2

Classification in Multi-Label Datasets

Aouatef Mahani

Abstract

Multi-label datasets contain several classes, where each class can have multiple values. They appear in several domains such as music categorization into emotions and directed marketing. In this chapter, we are interested in the most popular task of Data Mining, which is the classification, more precisely classification in multi-label datasets. To do this, we will present the different methods used to extract knowledge from these datasets. These methods are divided into two categories: problem transformation methods and algorithm adaptation ones. The methods of the first category transform multi-label classification problem into one or more single classification problems. While the methods of the second category extend a specific learning algorithm, in order to handle multi-label datasets directly. Also, we will present the different evaluation measures used to evaluate the quality of extracted knowledge.

Keywords: classification, instance, classifier, rank, label

1. Introduction

Classification is the most popular task in Data Mining. It consists of attribute to the appropriate class to an instance. There are several fields of classification that depend on the number of classes and the number of possible values of a class in a dataset. If a dataset contains a single class, which can have two values, then we speak about the classification in binary datasets. However, if the single class has more than two values, then we speak about the classification in multi-class datasets. In a case where a dataset contains several classes at a time, we speak about the classification in multi-label datasets.

Multi-label datasets appear in several applications such as text categorization [1], image annotation [2, 3], web advertising [4], and music categorization [5]. In these applications there are usually tens or hundreds of thousands of labels, while the number is still increasing. It is important to extract knowledge from these datasets to take decision. Consequently, the problem of classification in this kind of datasets is being an important problem in machine learning, and it has attracted the attention of many researchers.

For supervised learning algorithm from multi-label datasets, there are two major tasks: multi-label classification MLC and label ranking (LR) [6]. The first task is concerned with learning a model that outputs a bipartition of labels into relevant and irrelevant labels. The second task is concerned with learning model that outputs an ordering of the class labels according to their relevance.

For both tasks, the different approaches and techniques are proposed to deal with the classification in multi-label datasets divided into the two categories: problem transformation methods and algorithm adaptation methods [7]. In the first category, multi-label classification problem is transformed into one or more single classification problems. However, in the second one, the existing approaches are adapted to the studied problem.

This chapter is organized as follows: Section 2 presents the different notations used in the rest of chapter. In Section 3, we present at first the description measures of a multi-label dataset, and in the Section 4, we present the evaluation metrics used to evaluate the performances of the test dataset. In Section 5, we detail the different approaches and techniques used to deal with the problem of classification in multi-label datasets. Finally, in Section 6, we make our concluding remarks.

2. Notations

In the rest of this approach, we have used these notations:

D: is the concerned multi-label dataset.

N: is the size of multi-label dataset.

L: the set of labels.

Q: is the number of labels.

Y: is a set of labels, where Y is included in L with k = |Y|. It is known as k-label set.

$\overline{Y}$: is the complementary set of Y.

3. Description measures

In a multi-label dataset, the number of labels varies from one instance to another. For this reason, we can find some datasets that contain few labels compared with the total number of labels. This could be a parameter that influences the performance of different methods and approaches used to deal with the classification problem in multi-label databases. Therefore, a statistical analysis is necessary in order to have a description on a database [7, 8].

3.1 Label cardinality LC

LC indicates the average number of labels per instance (Eq. (1)).

$$LC = \frac{1}{N}\sum_{i=1}^{N} |y_i| \tag{1}$$

3.2 Label density LD

LD is the average number of labels divided by the overall number of labels Q (Eq. (2)).

$$LD = \frac{1}{N}\sum_{i=1}^{N}\frac{|y_i|}{Q} \quad (2)$$

3.3 Distinct label sets DL

DL counts the number of label sets that are unique across the total number of instances (Eq. (3)).

$$DL = |\{\exists x_i \in X \text{ and } Y_i \subseteq Y(x_i, Y_i) \in D \quad (3)$$

4. Evaluation measures

For classical classification, different performance measures have been proposed such as accuracy and coverage. However, the performance measures for classification in multi-label datasets are more complicated than single-class datasets. Consequently, a number of evaluation measures are proposed specifically to the multi-label datasets. These measures are categorized into two groups: example-based measures and label-based ones.

4.1 Example-based measures

These measures evaluate each instance of test dataset, and they return the mean value. They are also divided into two groups: prediction-based measures and ranking-based ones. The former measures use a learning system, and they are calculated based on the average difference of the actual and the predicted set of labels over all test instances. Whereas, the latter measures evaluate the label ranking quality depending on the scoring function f(.,.).

4.1.1 Prediction-based measures

Hamming Loss [9] represents the fraction of misclassified instances. If this measure is low, then the classifier has good performance (Eq. (4)).

$$\mathrm{H} = \frac{1}{\mathrm{N}} * \sum_{i=1}^{m} |\mathrm{Y_i}\Delta \mathrm{Z_i}| \text{ where : } \mathrm{Y_i}\Delta\mathrm{Z_i} = \mathrm{XOR}(\mathrm{Y_i}, \mathrm{Z_i}) \quad (4)$$

Classification Accuracy [10] represents the fraction of well-classified instances. It is a very strict as it requires the predicted set of labels to be an exact match of the true set of labels. It is also known as subset Accuracy [11] (Eq. (5)).

$$\text{Classification Accuracy} = \frac{1}{\mathrm{N}} * \sum_{i=1}^{\mathrm{N}} \mathrm{I}(\mathrm{Z_i} = \mathrm{Y_i}) \quad (5)$$

Where: $\mathrm{I}(\mathrm{Z_i} = \mathrm{Y_i})$=1 if $(\mathrm{Z_i} = \mathrm{Y_i})$ et 0 else.

Accuracy [12] represents the percentage of correctly predicted labels among all predicted and true labels (Eq. (6)).

$$\text{Accuracy} = \frac{1}{N} * \sum_{i=1}^{N} \frac{|Y_i \cap Z_i|}{|Y_i \cup Z_i|} \tag{6}$$

Precision represents the proportion of true positive predictions (Eq. (7)) [13].

$$\text{Precision} = \frac{1}{N} * \sum_{i=1}^{N} \frac{|Y_i \cap Z_i|}{|Z_i|} \tag{7}$$

Recall: estimates the proportion of true labels that have been predicted as positives (Eq. (8)) [13].

$$\text{Racall} = \frac{1}{N} * \sum_{i=1}^{N} \frac{|Y_i \cap Z_i|}{|Y_i|} \tag{8}$$

4.1.2 Ranking-based metrics [14]

Coverage error evaluates how many steps are needed, on average, to move down the ranked label list so as to cover all the relevant labels of the instance (Eq. (9)).

$$Coverage\ error = \frac{1}{N} * \sum_{i=1}^{N} max_{y \in Y_i} rank_f(X_i, y) - 1 \tag{9}$$

One-error computes how many times the top-ranked label is not in the true set of labels of the instance (Eq. (10)).

$$One - error = \frac{1}{N} * \sum_{i=1}^{N} \langle [\arg max_{y \in Y_i} f(X_i, y)] \notin Y_i \rangle \tag{10}$$

4.2 Label-based measures

In the aim to present the label measures, we compute the four components of the confusion matrix for each label y_i, which are TP_i, FP_i, TN_i, and FN_i that represent respectively true positive, false positive, true negative, and false negative (Eqs. (11)-(14) [15].

$$TP_i = |\{X_i\ where : y_i \in Y_i\ and\ y_i \in Z_i; 1 \le i \le N\}| \tag{11}$$

$$FP_i = |\{X_i\ where : y_i \notin Y_i\ and\ y_i \in Z_i; 1 \le i \le N\}| \tag{12}$$

$$TN_i = |\{X_i\ where : y_i \notin Y_i\ and\ y_i \notin Z_i; 1 \le i \le N\}| \tag{13}$$

$$FN_i = |\{X_i\ where : y_i \in Y_i\ and\ y_i \notin Z_i; 1 \le i \le N\}| \tag{14}$$

The label measures evaluate each label, and they return the average. The calculation of the average of all the labels can be achieved using two operations: macro-average and micro-average [16]. In macro-average, we calculate the performance measure of each label (Eqs. (15) and (16)), and then we take the average. On the other hand, in micro-average, we calculate the average performance measure of all the labels (Eqs. (17) and (18)).

Measures	*Macro-average measures*	*Micro-average measures*
Precision	$\frac{1}{L} * \sum_{i=1}^{L} \frac{TP_i}{TP_i + FP_i}$ (15)	$\frac{\sum_{i=1}^{L} TP_i}{\sum_{i=1}^{L} TP_i + \sum_{i=1}^{L} FP_i}$ (17)
Recall	$\frac{1}{L} * \sum_{i=1}^{L} \frac{TP_i}{TP_i + FN_i}$ (16)	$\frac{\sum_{i=1}^{L} TP_i}{\sum_{i=1}^{L} TP_i + \sum_{i=1}^{L} FN_i}$ (18)

5. Approaches and methods

The existing methods used to handle the classification problem in multi-label datasets are divided into two groups: problem transformation methods and algorithm adaptation methods.

5.1 Problem transformation method

This group transforms multi-label classification problem into one or more single classification problem [17].

5.1.1 Copy transformation method

This method [18] creates a single label dataset from original multi-label one. It replaces each multi-label instance with $|Y_i|$ labels by $|Y_i|$ instances. The variations of this method are dubbed copy-weight, select family of transformations, and ignore transformation. The first variation associates a weight to each produced instance. In the second one, for each set of created instances, only one instance is selected by applying the select max method that selects the most frequent instance, or the select min method that selects the least frequent instance, or select random one that selects an instance randomly. The last method deletes all multi-label instances.

5.1.2 Binary relevance (BR)

BR [17] is one of the most popular methods. It generates one dataset for each label where each dataset contains all instances, but with only one class, which may be positive or negative. For each instance of the ith dataset, if its set of labels contains the ith label, then its class is positive; otherwise its class is negative.

For each dataset, a classifier is generated. To classify a new instance, the BR method returns the union of all labels predicted by generated classifiers.

Although BR is a simple transformation method, it has been strongly criticized due to its incapacity of handling label dependency information [19].

5.1.3 Label power set (LP)

LP method [7] considers each set of labels of an instance as one class. For classifying a new instance, BR outputs the most probable class.

LP takes into account label dependence, but it has two drawbacks. First, the learning step becomes difficult when the number of label sets increases, especially when this number is exponential [20]. Second, the class imbalance problem can appear when there are some label sets that are represented by very few instances in the training dataset [20].

5.1.4 Random K-labelsets (RAKEL)

RAKEL [7] generates m Label Power set (LP) classifiers. To construct the LP classifier, we randomly select a k-labelset from L^k without replacement, and we build the appropriate training dataset. We note that the number of iterations m and the size of a label set k are the user-specified parameters. The different steps are detailed in this algorithm:

```
Input: training dataset D, set of labels L, parameters m and k.
Output: m classifiers and corresponding k-label sets Zi
Begin
 1. Construct the set R of all k-label sets
 2. for i:=1 to min(m, |L^k|) do
         2.1. Select randomly the k-label set Zi from R; R:=R/Zi
         2.2. Construct the corresponding training dataset Di:
              • Di:= Ø
              • For each instance (Xi,Si) from D do
                   o W:=Xi ∩ Zi
                   o If W = Ø, then replace Si by the empty class else replace Si by W
                   o Di:=Di U {(Xi,W)}
         2.3. Build the classifier Hi using Di
         End.
```

To classify a new instance, each classifier uses its corresponding k-label set as it is illustrated in this algorithm:

```
Input: new instance X, set of m k-label set Zi, L, m LP classifiers Hj and the threshold T.
Output: vector of predictions V
Begin
 1. for i:=1 to |L| do sumi:=0; votesi:=0
 2. for j:=1 to m do
      for each label li ϵ Zi do sumi:=sumi + Hj(X,li); votesi:=votesi + 1
 3. for i:=1 to |L| do
      Avgi:=sumi/votesi
      If (Avgi > T), then Vi:=1 else Vi:=0
 End.
```

5.1.5 Ranking by pair-wise comparison (RPC)

RPC [21] produces L*(L-1)/2 binary datasets from original dataset, one for each pair (l_i, l_j) with $1 \leq i < j \leq k$. Each dataset contains only instances that have the label l_i or l_j, but not both, and it is used to generate a binary classifier. To classify a new instance, each binary classifier outputs the labels, then the majority votes are applied for each label.

5.1.6 Calibrated label ranking (CLR)

CLR [22] is a technique that extends RPC by introducing a new virtual label. This latter is known as calibration label, and it is considered as a breaking point of the ranking that split the set of labels into two sets: relevant labels and irrelevant labels.

5.1.7 Classifier chain model (CC)

CC [19] produces L classifiers as Binary Relevance, but the actual classifier depends on previous one.

Example:

Attributes	*Label*	*Attributes*		*Label*	*Attributes*			*Label*	*Attributes*				*Label*
X_1	1	X_1	1	1	X_1	1	1	1	X_1	1	1	1	0
X_2	0	X_2	0	0	X_2	0	0	1	X_2	0	0	1	1
X_3	1	X_3	1	1	X_3	1	1	0	X_3	1	1	0	1
X_4	0	X_4	0	0	X_4	0	0	1	X_4	0	0	1	0
X_5	1	X_5	1	0	X_5	1	0	1	X_5	1	0	1	1

5.1.8 Ensemble of classifier chains (ECC)

This technique [23] uses classifier chains as a base classifier. It trains several CC classifiers using a standard bagging scheme. The produced binary models of each chain are ordered according to a random seed. Each model predicts different label sets. These predictions are summed per label so that each label receives a number of votes. A threshold is used to select the most popular labels that form the final predicted multi-label set.

5.1.9 Pruned sets (PS)

PS [24] consists of creating the new training dataset P from the original training dataset D by pruning infrequently label sets. This operation is controlled by a parameter p, which determines how often a label combination must occur for it not to be pruned. This algorithm summarizes this operation:

Input: the original dataset D and the parameter p.
Output: the pruned dataset P and the set of labels sets LC.
Begin

1. P:=Ø; LC:=Ø
2. for each instance (X_i,S_i) from D do
   If Si ϵ LC, then increment its frequency c by 1; else LC:=LC U $(S_i,1)$
3. for each instance (X_i,S_i) from D do
   - Use LC to retrieve the frequency of Si: (S_i,c)
   - If c > p, then P:=P U $\{(X_i,S_i)\}$ else (X_i,S_i) is considered as a pruned instance
4. for each pruned instance (X_i,S_i) do
   - Decompose *Si* into subsets s_{i0}, s_{i1}, … , s_{in} where each s_{ij} belongs to LC and its frequency c is >p
   - for each s_{ij} do form the new instance (d_i,s_{ij}); P:=P U $\{(d_i,s_{ij})\}$

End.

The pruned instances are reintroduced into the training in the form of new instances with smaller and more commonly found label sets. This allows the preservation of the instances and information about their label set. However, the size of training dataset is increased, and the average number of labels per instance becomes lower, which can in turn cause too few labels to be predicted at classification time.

5.1.10 Ensembles of pruned sets (EPS)

The PS method cannot create the new multi-label sets, which have not been seen in the training dataset. Consequently, it presents a problem when working with datasets where labelling is particularly irregular or complex. To solve this problem, an ensemble of PS [24] is proposed. The build phase of EPS is straightforward. Over m iterations, a subset of the training set is sampled and a PS classifier with relevant parameters is trained using this subset. For prediction, the threshold t is used, and different multi-label predictions are combined into a final prediction. This final label set may not have been known to any of the individual PS models, allowing greater classification potential.

5.1.11 Hierarchy of multilabel classifiers (HOMER)

HOMER [25] is an effective and computationally efficient for multi-label classification problem. Its principle consists of constructing a hierarchy of multi-label classifiers in the form of tree, following the divide-and-conquer strategy. Each deals with a much smaller set of labels compared with the set L. Each node of the tree contains a set of labels and the produced classifier, in which the root contains the set of all labels, and the leaves contain a single label. Each internal node contains the conjunction of the label sets of its children. The construction of this tree is done by following these steps:

1. The root contains all labels.

2. Train the classifier H_1 using all training dataset.

3. For each node n that contains more than a single label does.

    - Create k children.

    - Each child filters the training dataset of its parents by keeping instances that have at least one of its own labels.

    - Train the classifier H_n using the filtered dataset.

The question is how to distribute the labels of a node on k children?

For each child, the labels may be evenly distributed to k subsets in a way such that labels belonging to the same subset are as similar as possible. To do this, HOMER uses a balanced clustering algorithm, known as balanced k means.

5.2 Algorithm adaptation methods

5.2.1 Decision trees

The decision tree algorithm C4.5 [26] is efficient and robust for machine learning. It consists of constructing a tree top down in which the nodes contain the most suitable attributes. The selection of the suitable attribute is done by using the information gain (Eq. (19)), which is the difference between the entropy of the remaining instances in the training dataset and the weighted sum of the entropy of the subsets caused by partitioning on the values of that attribute.

$$information\ gain\ (D, A) = \text{entropty}(\text{D}) - \sum_{\text{v} \in \text{VA}} \frac{|\text{D}_\text{v}|}{|\text{D}|} * \text{entropy}(\text{D}_\text{v}) \tag{15}$$

Where: D is the training dataset, A is the considered attribute, VA is the set of possible values of the attribute A, D_v is the number of instances from the training dataset in which the value of the attribute A is v, and the entropy for a set of instances is defined in (Eq. (20)):

$$entropy\ (D) = -\sum_{\text{i}=1}^{\text{N}} \text{p}(\text{c}_\text{i}) * \log \text{p}(\text{c}_\text{i}) \tag{16}$$

Where: $p(c_i)$ is the probability (relative frequency) of class ci in this set.

The formula of the entropy is specific to a single class where the leaves contain one class. Therefore, C45 algorithm is the problem for multi-label datasets, and it is necessary to modify the formula of the entropy.

In [27], the learning process is accomplished by allowing multiple labels in the leaves of the tree. The formula for calculating entropy is modified for solving multi-label problems. The modified entropy sums the entropies for each individual class label (Eq. (21)).

$$entropy\ (D) = -\sum_{\text{i}=1}^{\text{N}} \text{p}(\text{c}_\text{i}) * \log \text{p}(\text{c}_\text{i}) + \text{q}(\text{c}_\text{i}) * \log \text{q}(\text{c}_\text{i}) \tag{17}$$

Where: $p(c_i)$ is the relative frequency of class label c_i, and $q(c_i)$ = 1- $p(c_i)$.

5.2.2 K-nearest neighbors KNN

Several methods exist based on KNN algorithm. ML-KNN [28] is the extension of KNN for classification problem in multi-label datasets. It consists of computing the prior and posterior probabilities to determine labels of a test instance. We introduce these notations before presenting ML-KNN.

- ***The category vector*** $\vec{y}$ of an instance X: is a vector of size L where $\vec{y}(l) = 1$ if $l \in Y$ and 0 otherwise.
- ***The K-nearest neighbors*** of X: N(X)
- ***The membership counting vector*** $\vec{C}$: to count the frequency of each label from N (X).
- ***The event*** H_1^l that the instance X has label l.
- ***The event*** H_0^l that X has not label l.
- ***The event*** E_j^l ($j \in \{0, 1, \ldots, k\}$) denote the event that, among the k nearest neighbors of X, there are exactly j instances, which have label l.

To classify the test instance T, we follow these steps:

- Compute the prior probability $P(\boldsymbol{H_1^l})$ of each label l using all the training dataset (Eq. (22)):

$$P(H_1^l) = s + \sum_{i=1}^{N} \vec{y}_{x_i}(l)/(s * 2 + N) \tag{18}$$

Where: N is the size of the training dataset, s is an input argument, which is a smoothing parameter controlling the strength of uniform prior.

- Determine the K-nearest neighbors of T.

- Compute the posterior probability $P\left(\boldsymbol{E_j^l}\backslash\boldsymbol{H_b^l}\right)$ for each label l and for each neighbor j (Eq. (23) and Eq. (24)):

$$P\left(E_j^l\backslash H_1^l\right) = s + c[j])/(s * (K+1))) + \sum_{p=0}^{K} C_1[p] \tag{19}$$

$$P\left(E_j^l\backslash H_0^l\right) = s + c'[j])/(s * (K+1))) + \sum_{p=0}^{K} C_2[p] \tag{20}$$

Where the vectors C_1 and C_2 are computed for each label and each instance.

- Compute the prediction using the posterior probabilities.

5.2.3 Support vector machine

The support vector machines (SVMs) have been extended to handle the multi-label problem. For example, Rank-SVM [29] defines a linear model based on a ranking system combined with a label set size predictor with the aim to minimize the ranking loss (Eq. (25)) and to maximize the margin.

$$\text{RLoss} = \frac{1}{N} + \sum_{i=1}^{N} \frac{1}{|Y_i| * |\overline{Y}_i|} * |R(x_i)| \tag{21}$$

Where $R(x_i) = \{(l_1, l_2) \in Y_i * \overline{Y}_i \mid f(x_i, l_1) \le f(x_i, l_2)\}$, $\overline{Y}_i$ denotes the complement of Y_i in Y, and f is the scoring function that gives a score for each label l interpreted as the probability that l is relevant.

5.2.4 Ensemble methods

AdaBoost.MH [30] is the extension of the AdaBoost algorithm, which is designed to minimize the hamming loss; for more details, see [31]. The minimization is done by decomposing the problem into k orthogonal binary classification problems.

AdaBoost.MR [30] is designed to find a hypothesis that ranks the correct labels at the top.

6. Conclusion

In this chapter, we have presented the classification problem in multi-label datasets, which is an important problem because these datasets appear in several domains. We have presented the description measures and the suitable metrics to evaluate the performances of the extracted knowledge. Then, we have reviewed the different approaches and methods used to deal, which are divided into two main categories: multi-label transformation methods and algorithm adaptation methods.

In future work, we are planning to present a state of the art about different approaches and techniques used to handle the classification problem in imbalanced multi-label datasets.

Author details

Aouatef Mahani
University of Science and Technology of Algiers Houari Boumediene USTHB, Algiers, Algeria

*Address all correspondence to: mahani.aouatef@gmail.com

References

[1] Dekel O, Shamir O. Multiclass-multilabel learning when the label set grows with the number of examples. In: Proceedings of 13th International Conference on the Artificial Intelligence and Statistics (AISTAT (2010)), 13-15 May 2010; Chia Laguna, Sardinia, Italy. 2010

[2] Deng J, Dong W, Socher R, Li Li-Jia, Li K, Li Fei-Fei. Imagenet: A large scale hierarchical image database. In: Proceedings of 2009 IEEE Conference on Computer Vision and Pattern Recognition (CVPR 2009), 20–25 June 2009; Miami, Florida, USA. 2009. pp. 248-255

[3] Nguyen CT, Zhan DC, Zhou ZH. Multi-modal image annotation with multi-instance multi-label LDA. In: Proceedings of the 23rd International Joint Conference on Artificial Intelligence (IJCAI'13), 3–9 August 2013; Beijing, China. 2013. pp. 1558-1564

[4] Eygelzimer A, Langford J, Lifshits Y, Sorkin G, Strehl A. Conditional probability tree estimation analysis and algorithm. In: Proceedings of the 25th Conference in Uncertainty in Artificial Intelligence (UAI 2009), 18–21 June 2009; Montreal, Canada. 2009. pp. 51-58

[5] Turnbull D, Barrington L, Torres D, Lanckriet G. Semantic annotation and retrieval of music and sound effects. IEEE Transactions on Audio, Speech and Language Processing. 2008;**16**:467-476. DOI: 10.1109/TASL.2007.913750

[6] Vembu S, Gärtner T. Label ranking algorithms: A survey. In: Furnkranz J, Hullermeier E, editors. Preference Learning Handbook. Berlin, Heidelberg: Springer; 2009. pp. 45-64. DOI: 10.1007/978-3-642-14125-6_3

[7] Tsoumakas G, Vlahavas I. Random k-labelsets: An ensemble method for multilabel classification. In: Proceedings of the 18th European Conference on Machine Learning (ECML 2007), 17–21 September 2007; Warsaw, Poland. 2007. pp. 406-417

[8] Tsoumakas G, Katakis I, Vlahavas I. Mining multi-label data. In: Maimon O, Rokach L, editors. Data Mining and Knowledge Discovery Handbook. 2nd ed. Boston, MA: Springer; 2010. pp. 667-685. DOI: 10.1007/978-0-387-09823-4_34

[9] Xia Y, Nie L, Zhang L, Yang Y, Hong R, Li X. Weakly supervised multilabel clustering and its applications in computer vision. IEEE Transaction on Cybernetics. 2016;**46**:3220-3232. DOI: 10.1109/TCYB.2015.2501385

[10] Zhu S, Ji X, Xu W, Gong Y. Multi-labelled classification using maximum entropy method. In: Proceedings of the 28th annual international ACM SIGIR conference on Research and development in Information Retrieval (SIGIR05); 15-19 August 2005; Salvador. 2005. pp. 274-281

[11] Ghamrawi N, McCallum A. Collective multi-label classification. In: Proceedings of the 2005 ACM Conference on Information and Knowledge Management (CIKM'05); Bremen. Germany. October 31 - November 5 2005. pp. 195-200

[12] Godbole S, Sarawagi S. Discriminative methods for multi-labeled classification. In: Proceedings of the 8th Pacific-Asia Conference on Knowledge Discovery and Data Mining (PAKDD 2004); 26–28 May 2004; Sydney, Australia. 2004. pp. 22-30

[13] Kanj S. Learning methods for multi-label classification [thesis]. The University of Technology of Compiègne. 2013

[14] Zhang ML, Zhou ZH. A review on multi-label learning algorithms. IEEE Transactions on Knowledge and Data Engineering. 2014;**26**:1819-1837. DOI: 10.1109/TKDE.2013.39

[15] Madjarov G, Kocev D, Gjorgjevikj D, Dzeroski S. An extensive experimental comparison of methods for multi-label learning. Pattern Recognition. 2012;**45**: 3084-3104. DOI: 10.1016/J.PATCOG.2012.03.004

[16] Yang Y. An evaluation of statistical approaches to text categorization. Information Retrieval. 1999;**1**:69-90. DOI: 10.1023/A:1009982220290

[17] Boutell M, Luo J, Shen X, Brown C. Learning multi-label scene classification. Pattern Recognition. 2004; **37**:1757-1771. DOI: 10.1016/j.patcog.2004.03.009

[18] Ganda D, Buch R. A survey on multi label classification. Recent trends in Programming Languages. 2018;**5**:19-23. DOI: 10.37591/RTPL

[19] Read J, Pfahringer B, Holmes G, Frank RE. Classifier chains for multi-label classification. Machine Learning. 2011;**85**:333-359. DOI: 10.1007/s10994-011-5256-5

[20] Cherman EA, Monard MC, Metz J. Multi-label problem transformation methods: A case study. Clei Electronic Journal. 2011;**14**:4. DOI: 10.19153/cleiej.14.1.4

[21] Hüllermeier E, Fürnkranz J, Cheng W, Brinker K. Label ranking by learning pairwise preferences. Artificial Intelligence. 2008;**172**:1897-1916. DOI: 10.1016/j.artint.2008.08.002

[22] Fürnkranz J, Hüllermeier E, Mencia EL, Brinker K. Multilabel classification via calibrated label ranking. Machine Learning. 2008;**73**:133-153. DOI: 10.1007/s10994-008-5064-8

[23] Read J, Pfahringer B, Holmes G, Frank E. Classifier chains for multi-label classification. In: Proceedings of the 13th European Conference on Principles and Practice of Knowledge Discovery in Databases and 20th European Conference on Machine Learning (ECML PKDD 2009), Part II, LNAI 5782; September 7-11 2009; Berlin. 2009. pp. 254-269

[24] Read J, Pfahringer B, Holmes G. Multi-label classification using ensembles of pruned sets. In Proceedings of the 2008 Eighth IEEE International Conference on Data Mining (ICDM'08); 15-19 December 2008; USA. 2008. pp. 995-1000

[25] Tsoumakas G, Katakis I, Vlahavas I. Effective and efficient multi-label classification in domains with large number of labels. In: Proceedings of the European Conference on Machine Learning and Principles and Practice of Knowledge Discovery in Databases (ECML PKDD) Workshop on Mining Multidimensional Data; Belgium 15-19 September 2008. pp. 30-44

[26] Lakshmi BN, Indumathi TS, Ravi N. A study on C.5 decision tree classification algorithm for risk predictions during pregnancy. Procedia Technology. 2016;**24**:1542-1549. DOI: 10.1016/J.PROTCY.2016.05.128

[27] Clare A, King RD. Knowledge discovery in multi-label phenotype data. In: Proceedings of the 5th European Conference on Principles of Knowledge Discovery in Databases (PKDD 2001);

3–7 September 2001; Freiburg, Baden-Württemberg, Germany. 2001. pp. 42-53

[28] Zhang ML, Zhou ZH. ML-kNN: A lazy learning approach to multi-label learning. Pattern Recognition. 2007;**40**: 2038-2048. DOI: 10.1016/j.patcog.2006.12.019

[29] Elisseeff A, Weston J. A Kernel method for multilabelled classification. In: Proceedings of the Annual ACM Conference on Research and Development in Information Retrieval (SIGIR05); 15-19 August 2005; Salvador. 2005. pp. 274-281

[30] Schapire RE, Singer Y. Boostexter: A boosting-based system for text categorization. Machine Learning. 2000; **39**:135-168. DOI: 10.1023/A:1007649029923

[31] Schapire RE, Singer Y. Improved boosting algorithms using confidence rated predictions. Machine Learning. 1999;**37**:297-336. DOI: 10.1023/A:1007614523901

Chapter 3

Research Trends in Library and Information Science in India during 2011 to 2018

Sutapa Paul and Bapan Kumar Maity

Abstract

Bibliometric analysis analyzes the research activity based on the doctoral theses that were submitted to the Shodhganga repository from 2011 to 2018 to determine the research trends in library and information science (LIS) in India. The findings demonstrate a shift in academic attention from fundamental LIS subjects to computer applications. Based on the data from Shodhganga, the INDCAT database, Vidyanidhi, and University News, this paper examined 612 PhD theses that were submitted at various universities in India during the course of these eight years.

Keywords: LIS research, research in LIS, PhD theses, research trend, LIS research-India

1. Introduction

The development of LIS as a field of study began with the basic skills of classification, cataloging, indexing, reference, bibliographical search, and professional values. Dr. S.R. Ranganathan played a vital role in the development of library education, libraries, and the library profession in India. Formal LIS education in India is nearly 100 years old.

In this study, an attempt has been made to assess the year-wise and subject-wise growth of new conceptual research in LIS, particularly during the period 2011–2018. This is a bibliometric study, which identifies the recent trend in LIS education in India by analyzing the doctoral thesis uploaded in the Shodhganga repository during the year 2011 to 2018.

Shodhganga is a reservoir of electronic theses, and dissertations are primary sources of research materials that originate from doctoral theses/dissertations submitted to the universities for the award of a PhD degree. It is mandatory to submit an electronic version of theses and dissertations by the researchers in universities to a national repository with an aim to facilitate open access to Indian theses and dissertations to the academic community worldwide.

2. Review of literature

There are so many studies on LIS research trend on the basis of articles published in national and international journals, but it has been found that only few studies have been done on the basis of PhD thesis, which is awarded by Indian universities. Maity and Hatuta's [1] study is an attempt to find out the research trends of library management in LIS and to quantitatively analyze the research activity in India based on doctoral theses that were already awarded in the period 1950–2012. Yadav and Gohain [2] have noted that the trends of LIS education in India under three important phases: genesis of LIS education; LIS education during 1947–2014; and emerging trends in the new millennium in order to cover the important milestones that occurred in Indian LIS education. The literature that informs the study came from a wide range of sources. The paper provides useful current information to LIS professionals, fraternities, researchers, students, and other stakeholders. According to Pandita and Singh [3], this study examines the research growth in the field of LIS at a global level for the period 2004–2013. The key areas analyzed in the study include research growth in LIS at the continental level, world's ten leading LIS research countries, citation analysis Hirsch Index (h-index), etc. The study is supported with empirical findings, for which data were retrieved from the SCI MagoJournal and Country Ranking, based on SCOPUS data source. The study is not exhaustive in nature, as it covers only those articles published in LIS journals indexed with this particular data source. According to Chatha [4], the aim of the study is to analyze the current research trends in LIS through journal articles. The study covers the number of publications, authors per year, cited items, country-wise distribution of articles, state-wise authorships, and current research areas in LIS. According to Dora and Kumar [5], the current study aims to understand the trends in LIS research during 2004–2015 by studying the published literature in Indian LIS journals. Singh [6] in his article study investigates and identifies research trends on LIS on the basis of PhD theses for the past nineteen years at the Department of Library & Information Science (DLIS), Aligarh Muslim University (AMU), India. Mondal and Roy's [7] paper is basically a bibliometric study based on 4993 citations from 53 PhD theses on economics under the University of Burdwan, West Bengal, India.

3. Aim of the study

The aim of the study is to analyze the current research trends in LIS through research theses in India. The bibliometric method is used to categorize the data. The data was gathered from the Shodhganga repository for the time span of 2011–2018. The study shows the current trend in LIS research publication, carried out to determine all the important points helpful for further research. The study covers the number of doctoral theses, the number of universities that are actively involved in research work, year-wise distribution of theses, state-wise contribution to research work, and current research areas in LIS.

All data are collected from Shodhganga repository; then, all the theses are individually analyzed by the titles and keywords and undergo a lot of tool subjects like classification, cataloging, academic library, management, budgeting, bibliometric study, information-seeking behavior, and so on. According to UGC-recommended syllabus, all these tool subjects are distributed under 13 broader subjects those are identified as main topics of research work in the LIS field. These broader subjects are:

1. Library and society
2. Library management
3. Classification/knowledge management
4. Cataloging/resource description
5. Information and communication technology
6. Information source, system, and center
7. Information literacy
8. Information-seeking behavior
9. Information economics and knowledge management
10. Content creation and technical writing
11. Digital library system and digital environment
12. Information system analysis and design
13. Library statistics and informatics

A reservoir of Indian theses is a digital repository of theses and dissertations submitted to Indian universities. It is maintained by INFLIBNET Center, which is an autonomous center of the University Grants Commission (UGC) of India and was initially located at the campus of Gujarat University, Ahmadabad. The repository has a collection of 210,661 theses and 6123 synopses.

It has been observed that "Shodhganga" is an important Indian initiative and will facilitate open access to Indian theses and dissertations to the world academic community. Online availability of electronic theses through centrally maintained digital repositories will not only ensure easy access and archiving of theses but will also help in raising the quality and standard of research.

4. Objectives of this study

The main objectives of this study are:

1. To find out the nature and direction of LIS research during 2011–2018
2. To know the yearly distribution of the subject of LIS research
3. To examine the research trend of LIS in India
4. To verify the diversity of current research in the LIS field

5. Limitation of the study

The study covers research works in LIS at the PhD level, which have been already awarded for doctoral degree by Indian universities. The period of coverage is from 2011 to 2018, a period of 8 years. It has considered only those titles of doctoral theses that appeared in the INFLIBNET-Indcat (http://incat.inflibnet.ac.in/indcat). Various universities that have an LIS department and have uploaded theses on INFLIBNET have been searched and data accrued year-wise.

6. Methodology

The main source of data collected is from university news, which periodically lists out the doctoral theses accepted by different universities and also uploaded in the thesis database in India and collected from Shodhganga. Data collected from these databases namely Shodhganga, INDCAT database, Vidyanidhi, University News data, and INFLIBNET (http://incat.inflibnet.ac.in/indcat) were searched with the key term "Library," "Library science", and "Library and Information science" to retrieve 612 records and were downloaded for the period of 2011 to 2018. All doctoral theses have been listed out and checked manually. After checking, 715 records were retained for the study. The whole work has been divided into three parts. The retrieved title has been grouped according to chronological growth, decade-wise growth, and university-wise distributions in one part. The broad and narrow subject distribution has also been made to find the actual subject trends in other parts. And the third part is state- and zone-wise distribution. The data has been analyzed quantitatively using statistical charts, diagrams, tables, etc.

Data interpretation under topic and year wise: Here, data are analyzed according to tools areas under 13 broad areas.

Table 1 shows 84 research studies, which were completed through the years 2011 to 2018. Out of these, 42.86% research studies were done on academic library, 20.24% on university library, 14.28% on special library, and 1.19% and 4.76% on public library and national library, respectively. Also 4.76% research studies were done on librarianship. So, the result shows that the maximum research studies were done on academic library.

Sl no.	Name of sub topic	2011	2012	2013	2014	2015	2016	2017	2018	Total	(%)
1	Academic Library	2	4	5	2	6	10	7	0	36	42.86
2	University Library	1	3	3	1	2	4	3	0	17	20.24
3	Special library	1	1	0	3	3	2	2	0	12	14.28
4	Public library	0	4	2	1	1	2	3	0	13	15.48
5	National library	0	0	0	0	1	0	0	0	1	1.19
6	Reference service	0	0	0	1	0	0	0	0	1	1.19
7	Librarian-ship	0	0	1	0	1	2	0	0	4	4.76
Total value		4	12	11	8	14	20	15	0	84	100
Percentage (%)		4.76	14.84	13.09	9.52	16.6	23.80	17.85	0	100	

Table 1.
Research on library society.

If we analyze the data according to the year, the highest research studies were done in 2016 (23.80%) and the second highest research in the year 2017 (17.85%). In the area of library and society, rest of the research percentage were 4.76%, 14.84%, 13.09%, 9.52%, and 16.6% in the year 2011, 2012, 2013, 2014, and 2015, respectively.

Figure 1 shows a graphical research on library society according to year.

Table 2 shows 104 research studies, which were completed through the years 2011 to 2018. Out of these, 56.73% research studies were done on management, 10.58% on quality management, 6.73% on organization, 5.73% research on book publishing / e- publishing and infrastructure, 2.88% on preservation and conservation, 4.80% on finance, and only 0.96% on both administration and acquisition. So, the result shows that the maximum research studies were done on management.

If we analyze the data according to year, the highest number of research studies were done in 2012 (18.25%) and second highest research in 2014 (16.35%). In the area of library management, rest of the research percentage were 12.5%, 15.38%, 15.38%, 10.5%, 7.69%, and 3.85% in the year 2011, 2013, 2015, 2016, 2017, and 2018, respectively.

Figure 2 shows research on library organization and administration based on the years.

Table 3 represents 13 (100%) research studies that were completed through the years 2011 to 2018. Out of these, 23.08% (total 3) research studies were done on classification, 69.23% (total 9) on collection development, and 7.69% (total 1) on information processing in the area of research on knowledge organization.

If we analyze the data according to the year, the highest number of research studies were done in 2012 (30.77%, where the total was 4), and the second highest was done in the years 2017 and 2018, where both percentage were same, that is, 15.38%. During the rest of the remaining years, 7.69% research studies were done in the area of knowledge management.

Figure 3 shows research on classification/knowledge organization based on the years.

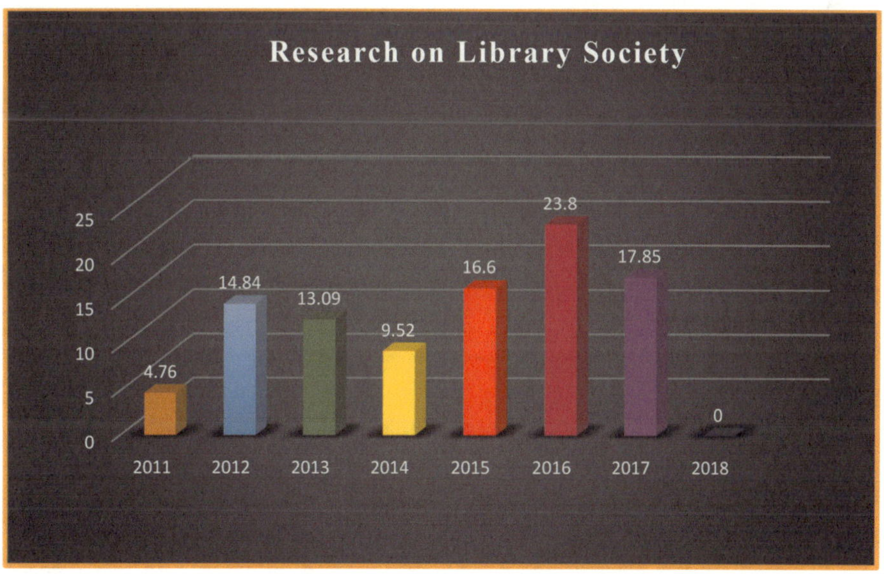

Figure 1.
Research on library society according to years.

Sl no.	Name of sub topic	2011	2012	2013	2014	2015	2016	2017	2018	Total	Percentage (%)
1	Management	9	9	10	9	11	8	2	1	59	56.73
2	Organization	1	1	1	2	1	0	0	1	7	6.73
3	Acquisition	0	1	0	0	0	0	0	0	1	0.96
4	Finance	0	3	2	0	0	0	0	0	5	4.80
5	Administration	0	1	0	0	0	0	0	0	1	0.96
6	Book publishing / e- publishing etc	1	1	1	3	0	0	0	0	6	5.78
7	Infrastructure	1	1	0	0	1	0	3	0	6	5.78
8	Preservation & conservation	1	0	1	0	1	0	0	0	3	2.88
9	Quality control	0	1	0	1	0	2	0	1	5	4.80
10	Quality Management	0	1	1	2	2	1	3	1	11	10.58
Total value		13	19	16	17	16	11	8	4	104	100
Percentage (%)		12.5	18.27	15.38	16.35	15.38	10.58	7.69	3.85	100	

Table 2.
Research on library management.

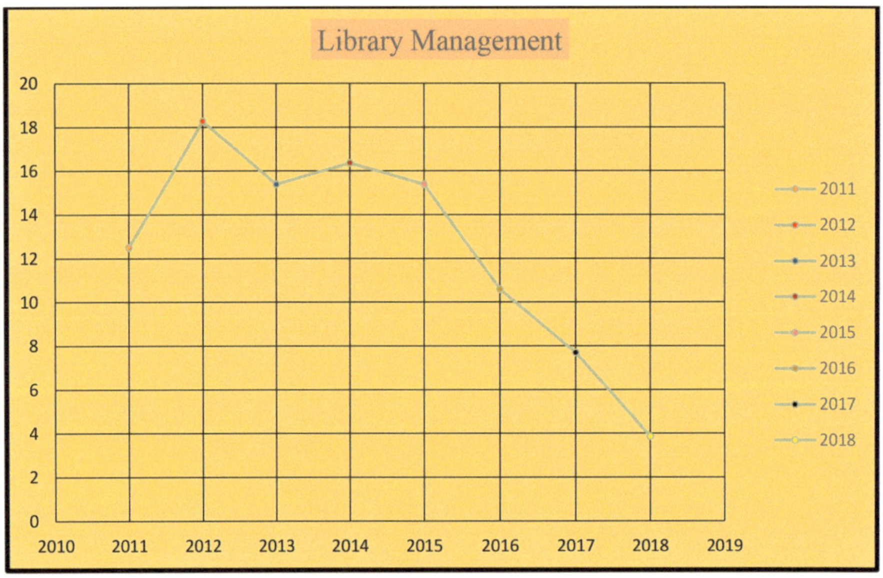

Figure 2.
Research on library organization & administration based on the years.

Table 4 shows that total 19 research studies were done on cataloging/resource description. Out of this, 26.31% research studies were completed on cataloging; 21.05% research studies were done on bibliography and indexing; and 15% research studies were completed on manuscript and control vocabulary. So, maximum research studies were done on cataloging.

Sl no.	Name of sub topic	2011	2012	2013	2014	2015	2016	2017	2018	Total	Percentage (%)
1	Classification	0	1	1	0	0	0	0	1	3	23.08
2	Collection development	1	2	1	1	1	1	1	1	9	69.23
3	Information Processing	0	1	0	0	0	0	0	0	1	7.69
Total value		1	4	2	1	1	1	1	2	13	100
Percentage (%)		7.69	30.77	15.38	7.69	7.69	7.69	7.69	15.38	100	

Table 3.
Research on classification/knowledge organization.

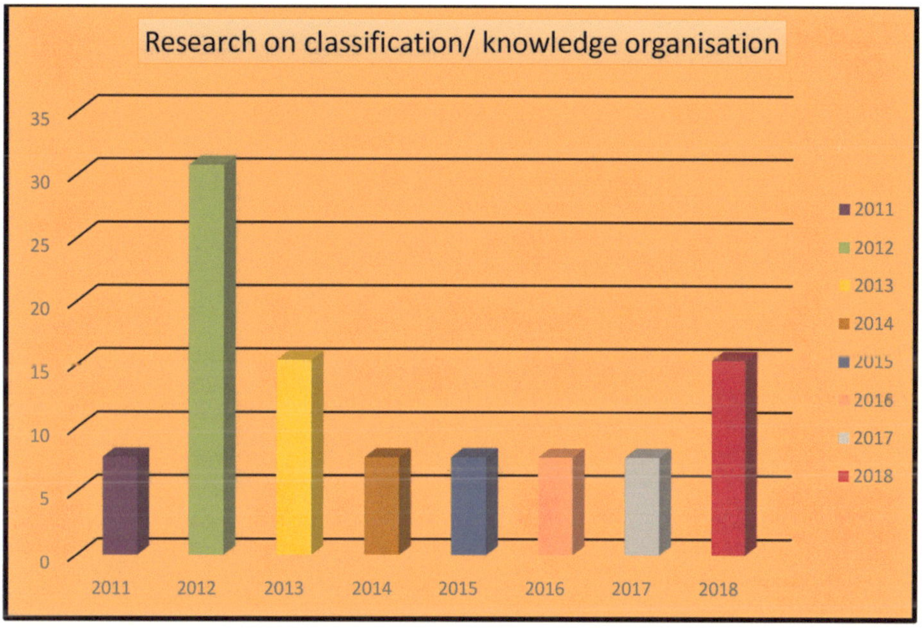

Figure 3.
Research on classification/knowledge organization based on the years.

On the basis of years (2011 to 2018), the analysis of data represents that 26.31% research studies were done in the year 2011, 21.05% in the year 2015 and 2018, 10.53% and 5.23% research studies were done in 2012 and 2016, respectively, in the research area of cataloging/resource description.

Figure 4 shows research on cataloging/resource description according to the year.

Table 5 shows 179 research studies, which were completed through the years 2011 to 2018. Out of these, 40.78% research studies were done on information technology, 12.85% on ICT, 11.73% on network, 11.17% on software, and 6.14% both internet and e-journals. Only 4.47%, 5.59%, and 1.12% research studies were done on database, automation, and electronic/print media, respectively. So, the result shows that the maximum research studies were done on information technology.

Sl no.	Name of sub topic	2011	2012	2013	2014	2015	2016	2017	2018	Total	Percentage (%)
1	Cataloging	0	0	3	0	2	0	0	0	5	26.31
2	Bibliography	3	0	0	1	0	0	0	0	4	21.05
3	Manuscript	0	0	0	0	0	0	0	3	3	15.79
4	Indexing	2	1	0	0	1	0	0	0	4	21.05
5	Control vocabulary	0	1	0	0	1	1	0	0	3	15.79
Total value		5	2	3	1	4	1	0	3	19	100
Percentage (%)		26.31	10.53	15.79	5.23	21.05	5.23	0	15.79	100	

Table 4.
Research on cataloging/research on resource description.

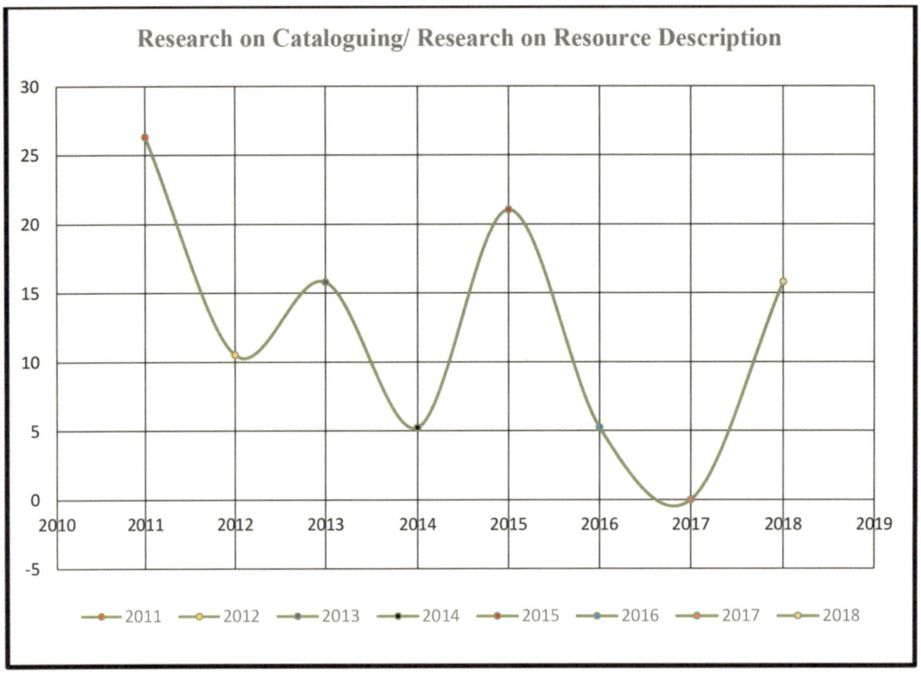

Figure 4.
Research on cataloging/resource description according to the year.

If we analyze the data according to the year, the highest number of research studies were done in 2012 (19.55%) and second highest research in 2013 (18.99%). In the area of information communication technology, rest of the research percentage were 9.50%, 11.17%, 13.96%, 11.17%, 11.73%, and 3.91% in the year 2011, 2014, 2015, 2016, 2017, and 2018, respectively.

Figure 5 shows research on information communication technology based on the years.

Table 6 reflects that total 29 (100%) research studies were done on information source, system, and center. Out of this, 13.79% (total 4) research studies were done on

Sl no.	Name of sub topic	2011	2012	2013	2014	2015	2016	2017	2018	Total	Percentage (%)
1	Database	0	4	1	1	1	1	0	0	8	4.47
2	Automation	2	3	1	3	1	0	0	0	10	5.59
3	Network	2	2	3	4	3	6	0	1	21	11.73
4	Software	7	3	4	1	1	1	3	0	20	11.17
5	Information Technology	3	16	11	7	13	9	10	4	73	40.78
6	ICT	0	5	6	2	3	0	6	1	23	12.85
7	Internet based	3	0	4	1	0	2	1	0	11	6.14
8	Electronic/ print media	0	1	0	0	1	0	0	0	2	1.12
9	E journal	0	1	4	1	2	1	1	1	11	6.14
	Total value	17	35	34	20	25	20	21	7	179	100
	Percentage (%)	9.50	19.55	18.99	11.17	13.96	11.17	11.73	3.91	100	

Table 5.
Research on information communication technology.

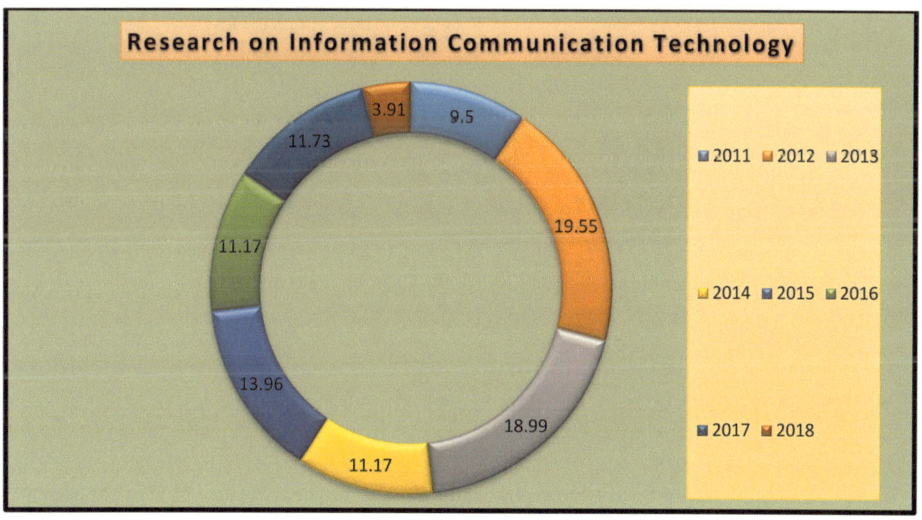

Figure 5.
Research on information communication technology based on the years.

information system, and 86.21% (total 25) research studies were done on information service/source.

Yearly distribution shows that 24.14% (7) doctoral theses were completed in the year 2015, which was the highest number of research on information source, system, and center during the years 2011 to 2018. The second highest was 20.69% (6) in the year 2012, whereas in the year 2018, no research was done on this topic. Rest of the research percentages with years were 13.79% (4) in 2011 and 2017, 6.9% (2) in 2013, and 10.34% (3) in 2014 and 2016.

Sl no.	Name of sub topic	2011	2012	2013	2014	2015	2016	2017	2018	Total	Percentage (%)
1	Information system	1	1	0	1	1	0	0	0	4	13.79
2	Information service/ source	3	5	2	2	6	3	4	0	25	86.21
Total value		4	6	2	3	7	3	4	0	29	100
Percentage (%)		13.79	20.69	6.9	10.34	24.14	10.34	13.79	0	100	

Table 6.
Research on information source, system, and services.

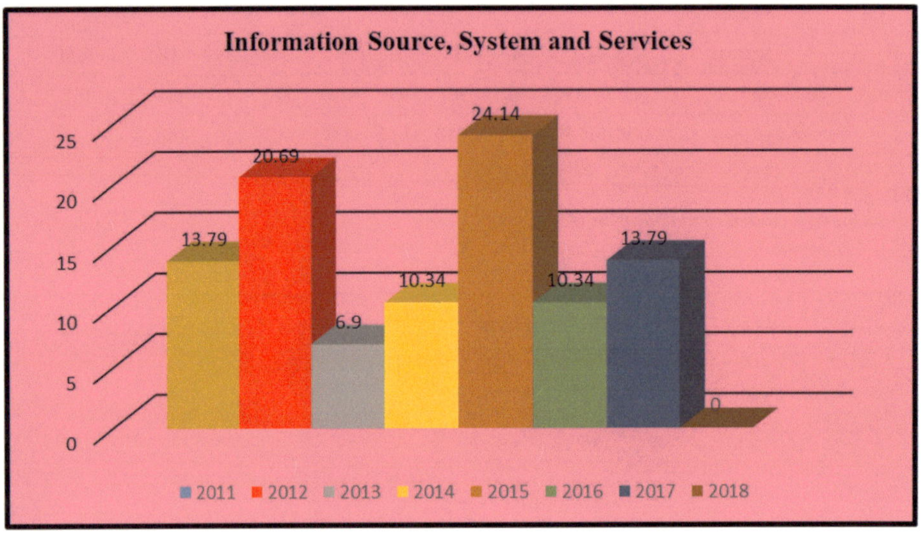

Figure 6.
Research on information source, system, and services according to the year.

Figure 6 shows research on information source, system, and services according to the year.

Table 7 shows that the total number of research on library classification in the year 2011–2018 was (21) 100%.

If we study according to the year, we can see the highest research studies were done in the year 2014 and 2017, where the percentage was same (23.81%; total 5 out of 21), and the second highest was in 2016 (19.05%; total 4), and 2015 had 14.28% (3)

Sl no.	Name of sub topic	2011	2012	2013	2014	2015	2016	2017	2018	Total	Percentage (%)
1	Information Literacy	1	1	1	5	3	4	5	1	21	100
Total value		1	1	1	5	3	4	5	1	21	100
Percentage (%)		4.76	4.76	4.76	23.81	14.28	19.05	23.81	4.76	100	

Table 7.
Research on information literacy.

research studies. Rest of the years were 2011, 2012, 2013, and 2018, which had done the same number of research studies on this topic, where the percentage were only 4.76 (total 1 out of 21).

Figure 7 shows research on information literacy based on the years.

Table 8 shows that total 86 research studies were done on seeking behavior and user approach in the years 2011–2018, where 41.86% were on information-seeking behavior and 47.67% were on user survey and user study. Only 10.47% research studies were done on reading habit. So, maximum research studies were done on user survey and user study.

If we study according to the year, we can see that the highest research studies were done in the year 2015, where the percentage was the same (17.44%; total 15 out of 86), second highest was in 2012 (16.28%), and 2013 and 2017 both had equal percentage, that is, 15.12% (total 12). Rest of the years were 2011, 2014, 2016, and 2018, which had done 10.46%, 13.95%, 9.30%, and 2.32%, respectively.

Figure 8 shows research on information-seeking behavior and user Approach.

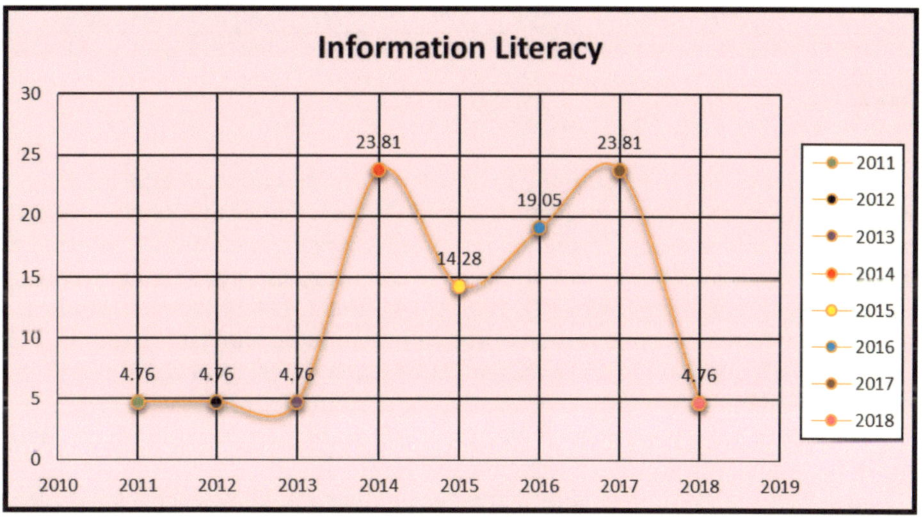

Figure 7.
Research on information literacy based on the years.

Sl no.	Name of sub topic	2011	2012	2013	2014	2015	2016	2017	2018	Total	Percentage (%)
1	Information seeking behavior	5	7	5	7	5	3	3	1	36	41.86
2	User survey and User study	2	7	7	4	9	5	6	1	41	47.67
3	Reading habit	2	0	1	1	1	0	4	0	9	10.47
Total value		9	14	13	12	15	8	13	2	86	100
Percentage (%)		10.46	16.28	15.12	13.95	17.44	9.30	15.12	2.32	100	

Table 8.
Research on information-seeking behavior and user approach.

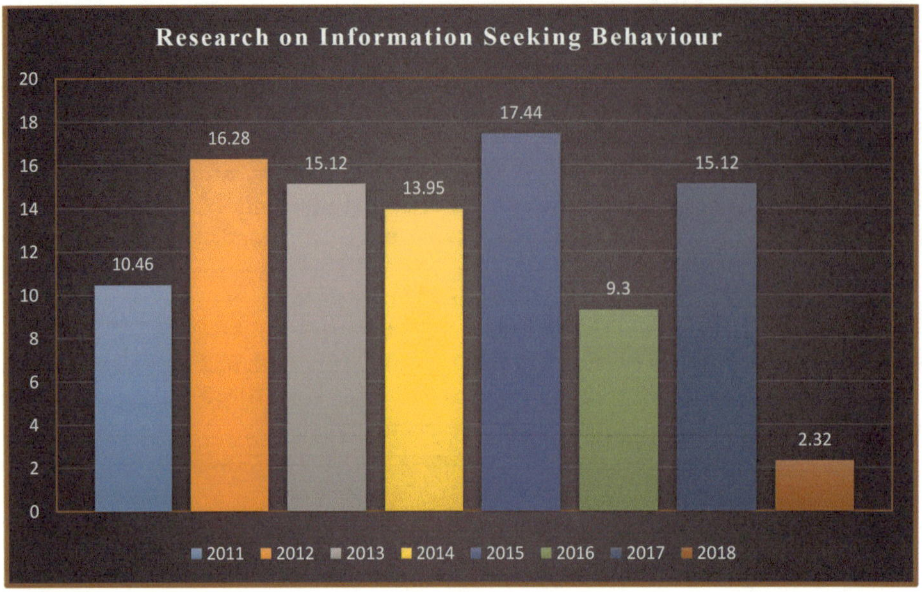

Figure 8.
Research on information-seeking behavior and user approach.

Table 9 shows that the total number of research studies done on information economics and knowledge management was 37 (100%). Out of this, 64.86% research studies were completed on information resource, where the total was 24, which is the highest among all, 16.21% research studies were done on resource sharing, where the total was 6, and remaining 5.40% (2), 10.81% (4), and 2.70% (1) research studies were done on marketing, web resource, and information dissemination, respectively.

Yearly distribution of library management research theses from 2011 to 2018 shows that among the 37 (100%) theses, the highest number of research studies were done in the year 2014; percentage is 29.7 (14). Second highest research studies were done in the years 2015 and 2017; percentage is 16.21 (6).

Figure 9 shows research on information economics and knowledge management based on the years.

Sl no.	Name of sub topic	2011	2012	2013	2014	2015	2016	2017	2018	Total	Percentage (%)
1	Marketing	1	1	0	0	0	0	0	0	2	5.40
2	Information resource	1	0	2	8	4	3	5	1	24	64.86
3	Web Resource	0	0	1	2	0	1	0	0	4	10.81
4	Information dissemination	0	0	0	0	1	0	0	0	1	2.70
5	Resource sharing	0	1	1	1	1	1	1	0	6	16.21
Total value		2	2	4	11	6	5	6	1	37	100
Percentage (%)		5.40	5.40	10.81	29.73	16.21	13.51	16.21	2.70	100	

Table 9.
Research on information economics and knowledge management.

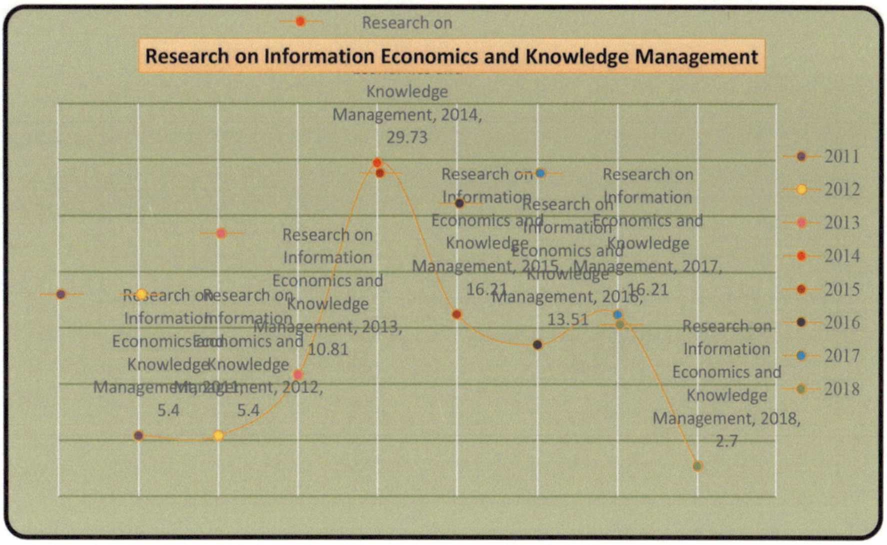

Figure 9.
Research on information economics and knowledge management based on the years.

Table 10 describes that total 10 (100%) research studies were completed on content creation and technical writing. Out of this, 50% research studies were done on open access, 30% research studies on e-learning, and 20% research studies on content management.

Yearly distribution shows that the highest research studies were done in the year 2015, which is 40%, and remaining 20%, 30%, and 10% research studies were completed in the years 2013, 2014, and 2016, respectively. 2011, 2012, 2017, and 2018 had no research in content creation and technical writing.

Figure 10 shows research on content creation and technical writing based on the years.

Table 11 shows 58 (100%) research studies that were completed on digital library system and digital environment. Out of this, 41.38% (24) research studies were done on the digital library system, which had the highest score; the second highest research score 34.48% (20) was on e-resource, 13.79% (8) on e-journal, and 5.17% (3) research studies were completed on both metadata harvesting and digital repository. So, the result shows that the maximum research studies were done on the digital library system.

Sl no.	Name of sub topic	2011	2012	2013	2014	2015	2016	2017	2018	Total	Percentage (%)
1	Open access	0	0	1	1	3	0	0	0	5	50
2	E-learning	0	0	0	1	1	1	0	0	3	30
3	Content management	0	0	1	1	0	0	0	0	2	20
Total value		0	0	2	3	4	1	0	0	10	100
Percentage (%)		0	0	20	30	40	10	0	0	100	

Table 10.
Research on content creation and technical writing.

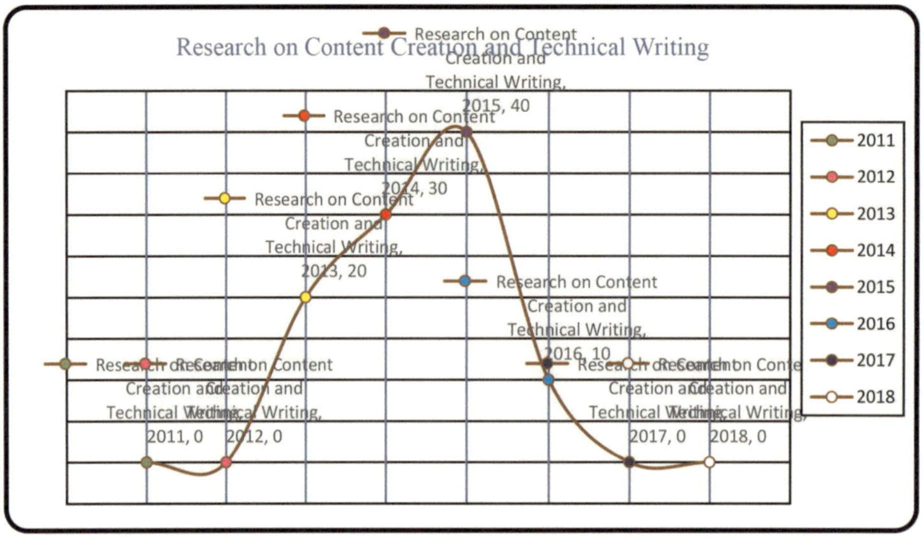

Figure 10.
Research on content creation and technical writing based on the years.

Sl no.	Name of sub topic	2011	2012	2013	2014	2015	2016	2017	2018	Total	Percentage (%)
1	Digital library system & Digital Environment	3	4	1	2	4	4	3	3	24	41.38
2	Metadata harvesting	2	1	0	0	0	0	0	0	3	5.17
3	Digital repository	0	1	1	0	0	0	1	0	3	5.17
4	E- resource	1	3	0	1	6	0	8	1	20	34.48
5	E- journal	0	1	4	1	2	0	0	0	8	13.79
	Total value	6	10	6	4	12	4	12	4	58	100
	Percentage (%)	10.34	17.24	10.34	6.89	20.69	6.89	20.69	6.89	100	

Table 11.
Research on digital library system and digital environment.

If we analyze the data according to the year, highest research studies were done in 2015 and 2017, both (6.89%). In the area of digital library system and digital environment, rest of the research percentage were 17.24% in 2012, 10.34% in both 2011 and 2013, 6.89% in 2014, 2016, and 2018.

Figure 11 shows research on digital library system and digital environment based on the years.

Table 12 shows that total 40 (100%) research studies were completed on information system analysis and design. Out of this, highest research studies were done on library professional (40%). Second highest was on quality management (30%) and rest of 15% (6), 12.5% (5), and 2.5% (1) research studies were completed on consortia, performance evaluation, and Thesaurus construction, respectively.

Yearly distribution shows that highest research studies were in the year 2012 and 2013, the score was same 17.5% (7); 2014 and 2015 also had done same percentages of

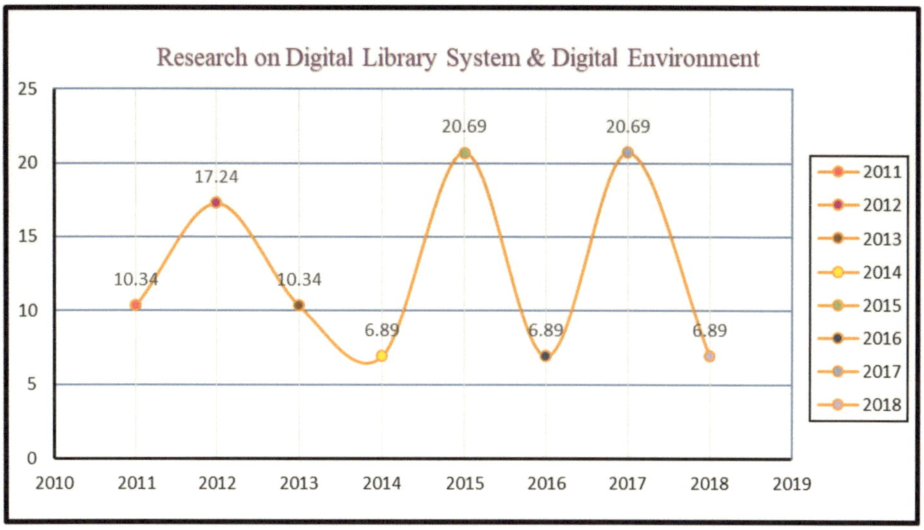

Figure 11.
Research on digital library system and digital environment based on years.

Sl no.	Name of sub topic	2011	2012	2013	2014	2015	2016	2017	2018	Total	Percentage (%)
1	Quality Management	1	1	1	2	2	1	3	1	12	30
2	Consortia	0	1	3	0	1	0	1	0	6	15
3	Performance evaluation	0	1	2	0	1	0	1	0	5	12.5
4	Library Professional	1	4	1	3	1	4	1	1	16	40
5	Thesaurus construction	0	0	0	0	0	1	0	0	1	2.5
Total value		2	7	7	5	5	6	6	2	40	100
Percentage (%)		5	17.5	17.5	12.5	12.5	15	15	5	100	

Table 12.
Research on information system analysis and design.

research studies, which was 12.5 (5). 2016 and 2017 also had same 15% (6), and 5% (2) research studies were done in the year 2011 and 2018.

Figure 12 shows research on information system analysis and design according to years.

Table 13 reflects that total 86 (100%) number of research studies were done on library statistic and informatics. Out of this, 61.17% (total 52) research studies were done on scientometric, which got highest number among these, and second highest research was 16.47% (total 14) done on citation analysis. Rest of 7.05% (total 6), 2.35% (total 2), 5.88% (total 5), and 8.14% (total 7) research studies were completed on bibliometric, webometric, hypotheses testing, and content analysis, respectively.

Yearly distribution shows that 20% (total 17) doctoral thesis were completed in the year 2014, which was the highest number of research on library statistic and informatics during the year 2011 to 2018. Second highest was 17.47% (total 17) in the year

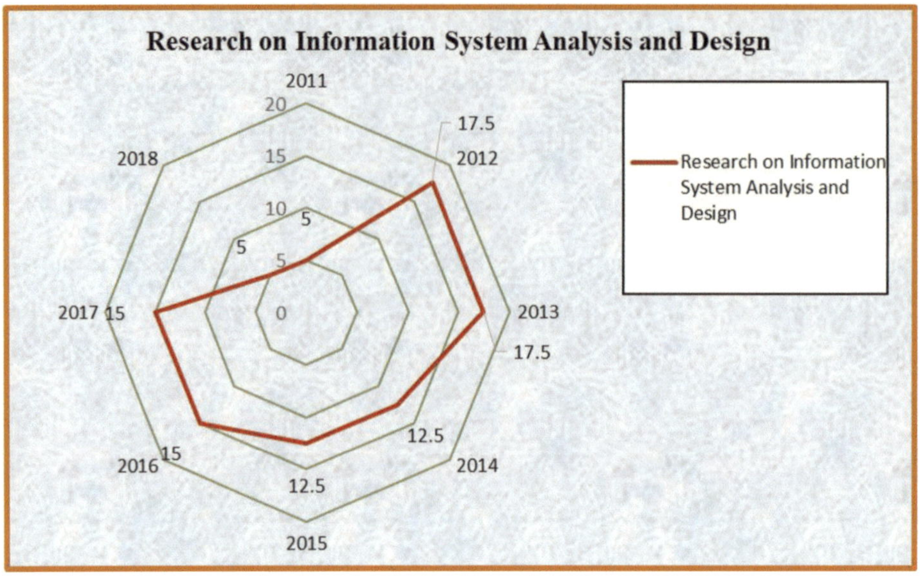

Figure 12.
Research on information system analysis and design according to years.

Sl no.	Name of sub topic	2011	2012	2013	2014	2015	2016	2017	2018	Total	Percentage (%)
1	Bibliometrics	2	1	0	1	0	1	1	0	6	7.05
2	Citation analysis	1	1	1	3	3	1	4	0	14	16.47
3	Webometrics	1	0	0	0	0	1	0	0	2	2.35
4	Hypothesis testing	0	2	1	2	0	0	0	0	5	5.88
5	Scientometric	4	8	3	9	14	9	4	1	52	61.17
6	Content analysis	0	0	1	0	0	2	3	1	7	8.14
Total value		8	12	6	15	17	14	12	2	86	100
Percentage (%)		9.41	14.11	7.05	17.64	20	16.47	13.95	2.35	100	

Table 13.
Research on library statistics and informatics.

2014. Rest of the research percentages with years were 9.41% (8) in 2011, 14.11% (12) in 2012, 7.05% (6) in 2013, 16.47% (14) in 2016, 13.95% (12) in 2017, and 2.35% (2) in the year 2018, respectively.

Data analysis and interpretation under broad subject with years: Here, data is represented and analyzed according to 13 broad subjects included with yearly distributions.

Figure 13 shows research on library statistic and informatics based on the year.

This table **(Table 14)** represents the research studies under broad subject. Here, we can see that the highest research studies were done under the subject "Information Communication Technology", the total number of research 179 (25.03%) out of 715 (100%). Second highest research studies were done in "Library Management"

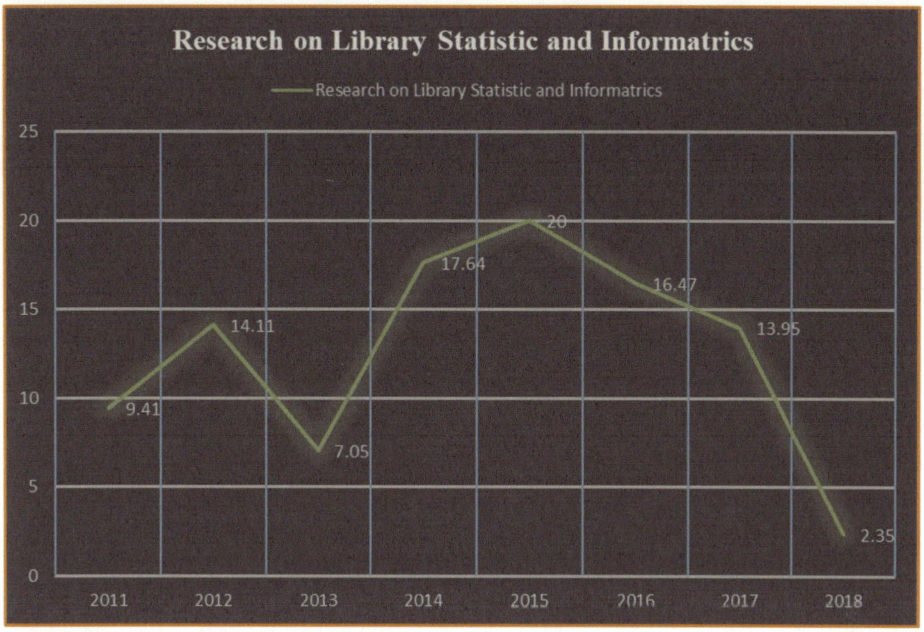

Figure 13.
Research on library statistics and informatics based on the year.

SL. No	Name of sub topic	2011	2012	2013	2014	2015	2016	2017	2018	Total	Percentage (%)
1	Library and Society	4	12	11	8	14	20	15	0	84	11.75
2	Library Management	13	19	16	17	16	11	8	4	104	14.54
3	Research on Classification/ Knowledge Organization	1	4	2	1	1	1	1	2	13	1.82
4	Cataloging/ Resource description	5	2	3	1	4	1	0	3	19	2.66
5	Information Communication Technology	17	35	34	20	25	20	21	7	179	25.03
6	Information Source, System and Services	4	6	2	3	7	3	4	0	29	4.06
7	Information Literacy	1	1	1	5	3	4	5	1	21	2.94
8	Information Seeking Behavior	5	7	5	7	5	3	3	1	36	5.03
9	Information Economics and Knowledge Management	2	2	4	11	6	5	6	1	37	5.17
10	Content Creation and Technical Writing	0	0	2	3	4	1	0	0	10	1.4

SL. No	Name of sub topic	2011	2012	2013	2014	2015	2016	2017	2018	Total	Percentage (%)
11	Digital Library System & Digital Environment	6	10	6	4	12	4	12	4	58	8.11
12	Information System Analysis and Design	2	7	7	5	5	6	6	2	40	5.59
13	Library Statistic and Informatics	8	12	6	15	17	14	11	2	85	11.89
Total value		68	117	99	100	119	93	92	27	715	100
Percentage (%)		9.51	16.36	13.85	13.99	16.64	13	12.88	3.77	100	

Table 14.
Data analysis under broad subject.

(14.54%) total were 104, where 85 (11.89%) research studies were on "Library Statistic and Informatics". "Library and Society" accrue 84 (11.75%) research studies. Next, "Digital Library System & Digital Environment", "Information System Analysis and Design", and "Information Economics and Knowledge Management" completed 58 (8.11%), 40 (5.59%), and 37 (15.17%) research studies, respectively. Rest of the subjects are "Research on Classification/ Knowledge", "Organization Cataloging/ Resource description", "Information Source", "System and Services", "Information Literacy", and "Information Seeking Behavior" completed 1 to 36 (6% to <2%) research studies in LIS field during 2011 to 2018.

According to years, the maximum number of research studies were in the years 2015 and 2012; the total were 119 (16.64%) and 117 (16.36%), respectively, where medium number of research studies were in the years 2014–100 (13.99%), 2013–2099 (13.85%), 2016–2093 (13%), 2017–2092 (12.88%), and 2011–2068 (9.51%), where lowest number of research studies were in the year 2018–2027 (3.77%) on the basis of theses uploaded in Shodhganga.

Figure 14 shows research studies under broad subject, and also **Figure 15** shows yearly distribution of research studies under broad subjects.

7. Findings of broad subject with years

According to the UGC recommended syllabus of LIS, all data which was collecting from INFLIBNET are content analysis and made through title and keyword and categorized in to 20 broad subjects, which is already discussed in Chapter 1. and Chapter 3. To identify the research trend in the area of subjects in LIS field, it will have to focus on findings. These are:

- After analyzing the data through the year 2011 to 2018, we can see that maximum research studies were done on **"Information Communication Technology"**, which got highest number of research studies, that is, 179 (25.03%) out of 715.

- **"Library Management"** got second position. The total number of research studies done on this subject was 104 (14.75%).

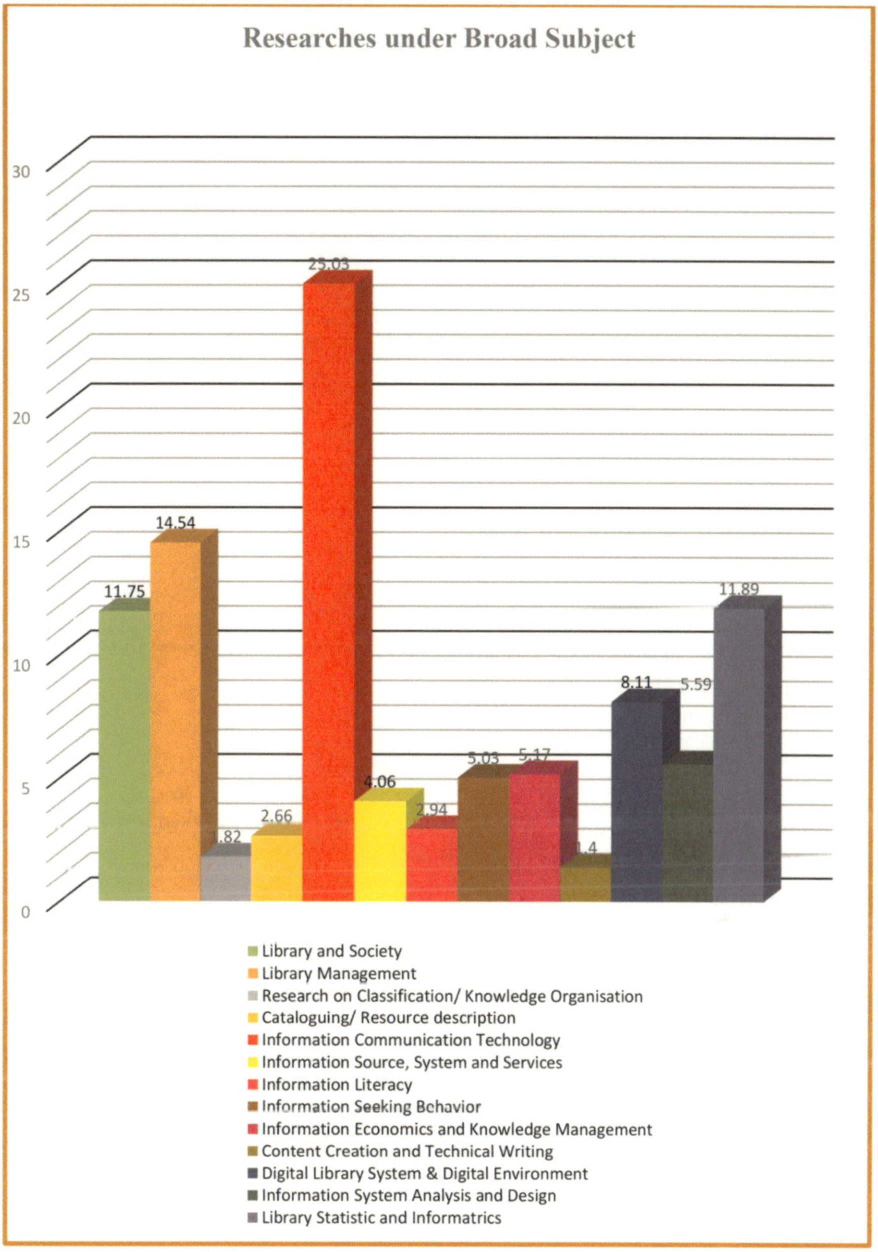

Figure 14.
Research studies under broad subject.

- **"Library Statistic and Informatics"**, **"Library Society"** also had good number of research studies, which accrue 85 (11.89%) and 84 (11.75%), respectively.

- Next "Digital Library System & Digital Environment" and "Information System Analysis and Design" completed 58 (8.11%) and 40 (5.59%), respectively, which is also good.

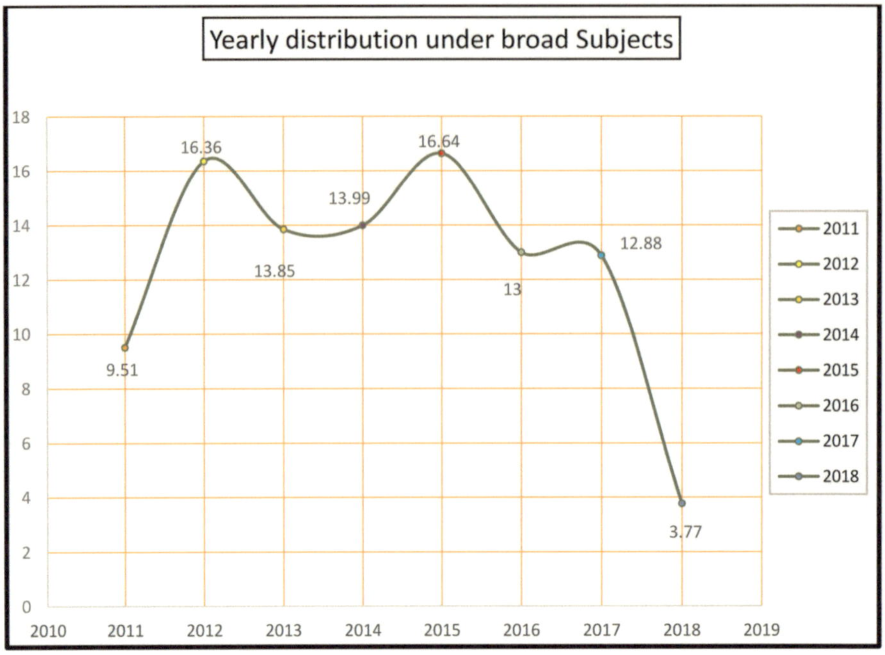

Figure 15.
Yearly distribution of research studies under broad subjects.

- But research studies on these subjects, that is, "Information Economics and Knowledge Management", "Information Seeking Behavior", and "Information Source, System, and Services" is medium.
- Lowest number of research studies were done under content creation and technical writing (1.4%), classification/ knowledge organization (1.82%), cataloging/resource description (2.66%), and information literacy (2.94%).
- On the basis of theses, which were uploaded to Shodhganga during 2011–2018, the yearly distribution showed that maximum number of research studies was in the years 2015 and 2012; the total was 119 (16.64%) and 117 (16.36%), respectively. The lowest research studies were done in the year 2018 (3.77%).

8. Conclusions

The current study shows that most of the doctoral theses that have been submitted in LIS discipline in India gradually come out from the core area of LIS fields like classification, cataloging, knowledge management, etc. The research trend is growing forward to the scholarly communication, IT based, management oriented, and in academic field also. Most of the research studies are being done on library statistics like bibliometric, scientrometric, webometrics, citation analysis, etc. Also, research on library society, information technology, application of information technology, digital library system and digital environment, information system analysis and design, and user approach is among the modern area of research of LIS field.

DOI: http://dx.doi.org/10.5772/intechopen.109092

Author details

Sutapa Paul[1] and Bapan Kumar Maity[2]*

1 Kothari Medical Centre, Kolkata, India

2 Sammilani Mahavidyalaya, Kolkata, India

*Address all correspondence to: bkmaity84@gmail.com

References

[1] Maity BK, Hatua SR. Research trends of library management in LIS in India since 1950–2012. Scientometrics. 2015; **105**:337-346. DOI: 10.1007/s11192-015-1673-8

[2] Yadav AK, Gohain RR. Growth and development of LIS education in India. SRELS Journal of information Management. 2015;**52**(6):403-414

[3] Pandita R, Singh S. Research growth in LIS during last decade: A study. Library Review. 2015;**64**:514-532. DOI: 10.1108/LR-04-2015-0037

[4] Chatha HS. Research trends in LIS: A bibliometric study of information studies journals during 2011-2015. International Research Journal of Management Science & Technology. 2016;**7**(7):03-07. Retrieved February 02, 2019, from: http://www.irjmst.com

[5] Dora M, Kumar HA. An empirical analysis of the research trends in the field of library and information science in India –2004-2015. COLLNET Journal of Scientometrics and Information Management. 2017;**11**(2):661-678. DOI: 10.1080/09737766.2017.1317959

[6] Singh RK, Singh AP, Singh AK. Research trends in field of library and information science: A case study of DLIS Aligarh Muslim University through Shodhganga repository. Library Progress (International). 2018;**35**:158-165. DOI: 10.5958/2320-317X.2018.00016.8

[7] Mondal S, Roy BK. Research trends of sub-subjects on economics: A bibliometric study. Library Philosophy and Practice. 2018. Retrieved from: https://search.proquest.com/docview/2165581881?accountid=16284

Chapter 4

Network Analysis in the Information Systems Management: Implications for a Transdisciplinary Approach

Massimo Bianchi

Abstract

The Network Analysis in organizations made in last years some meaningful results owing progress in technology and in the approach to organizational networks. The chapter enhances the need to transfer some results of network analysis from management to the analysis of technical networks. Coming from results in the studies, theoretical and empirical, on business ties and on strong and weak ties connected to the mechanism of control, the chapter proposes a transdisciplinary approach to interpret the differences and the evolution of the types of networks through cycles of simplification and complexification of the control systems. Particularly, as results are connected to the adequacy of control tools, it is relevant to consider managerial concepts, such as the span of control defined as the number of subordinates of a hierarchical position, and the connected capability of networks to maintain control, particularly when the system is wide and highly interconnected.

Keywords: control systems, network analysis, network management, span of control, transdisciplinary approach

1. Introduction

According to transdisciplinary approach, the introduction of new perspectives needs the definition of the scenario in which hypotheses are located and induce new tools of analysis.

Continuous progress in technology and organizational work approach led to an overlap of the concepts of *network* and *organization*. The network seems to be a modern expression of the organization, and the organization increasingly takes on the nature of a network.

The purpose of this chapter is to discuss the results of the network analysis of organizations and the relevance of the organizational concept in the meaningful understanding of networks and their control systems, which is not limited to a pure application of increasingly sophisticated algorithms but seems to become distant from the practice of organizations [1].

The perspective assumed is transdisciplinary, according to the distinction claimed by some authors, regarding its difference from multi and interdisciplinary perspectives [2]; as it implies the use of a model having the organizational matter as basic but with a reciprocal influence on information knowledge about networks.

Network analysis, in the last few years, has undergone a hyperbolic development of digitization and the creation of sophisticated applications [3]. However, this development did not consider the change in organizational conditions in which the situations are analyzed and enormously implemented in the last decades with the complexity of networks [4].

In other words, the hypothesis is that the qualitative side of this evolution has been brought about by the evolution of the network analysis that does not consider the evolution of the networks themselves as products of organizations, their feedback regulations [5], or the myopia of individuals in considering network performances [6]. Consequently, the field of analysis to which we apply is not thought to have changed, while practitioners observe this phenomenon and recognize that it is progressing rapidly [7], in a parallel way with the implementation of transdisciplinary research process [8].

The qualitative aspects of this evolution are striking compared to the quantitative aspects derived from the hyperbolic growth of parameters and indices used to measure the performance of networks. This evolution foreshadows not a simple adaptation of the reference models but a real phase shift [9].

This last step, still in progress [10], represents the limit of the complexification of models built to interpret situations and prepares for the next advancement by exceeding the threshold between two different phases and collapsing pre-existing models in favor of a newer, simpler model [11].

With the organizational evolution, this cycle of simplification that increasingly moves toward complexification and vice-versa, represents the organizational engine of the evolution of the analyzed social and corporate entities, according to the Kuhn hypothesis on scientific revolutions [12, 13].

Far from providing a definitive answer, this chapter represents an invitation to orient research toward the analysis of the evolution of information systems from an organizational viewpoint, considering strictly technical parameters, and broadening the consideration to organizational aspects in a transdisciplinary approach [14].

Different approaches were devised to give a transdisciplinary perspective to the theme of network analysis, already extensively treated in a traditional key. One of these, considered appropriate, is to define the general reference model for which to carry out the analysis [15].

In the specific case, the network analysis can be referred to as a model to which various interacting aspects refer, ranging from the classification of the types of networks to the simplification/complexification cycles of production processes, attention paid to details or to the general framework, and definition of intrinsic/extrinsic quality to end up in the control area that the network structure is able to express (**Figure 1**).

2. The first step of the analysis: simplification and complexification cycles

Considering the network to be a universally diffused model structure, we examine the main typologies of macro structures that make more evidently the form of governance (**Figure 2**).

In a basic evolutionary model, network macro structures span from the largest organizations (measured by the number of employed people), structured as

Topic	Definition
Network typology	Structural configurations linked to different dimensions and forms of network governance
Simplification/Complexification	Reduction / expansion of the number of indices used to measure network performance
Prevailing attention to Details/ General framework	Governance of the network oriented to details or to an overview of the system
Intrinsic/Extrinsic quality	Concept of product quality linked to the specific characteristics of the product or to the network that produces it
Restricted/Wide Span of control	Small or large number of positions dependent on a single superior
Renewal system of elements and ties	The continuous renewal of network due to acquisition or loosing of elements and ties

Figure 1.
Components of transdisciplinary model for network analysis.

autocratic networks, to smaller organizations that can employ as few as a single person managing activities.

Once established, the decentralized network produces an increase in networking among its units and, in the next phase, the emergence of a leader organization, which centralizes some strategic functions dominating the network. This step just precedes the creation of an integrated network, restarting the cycle.

The existence of different network typologies is accepted by organizational theory [16]. The novelty comes from the studies on the motivation for these differences and of their creation.

Starting from these organizational typologies, (**Figure 2**) a hypothesis on evolutionary model of macrostructures takes as an engine of change the continuous simplification or complexification cycle, which leads to the intrinsic quality of the process/ product from one side and an extrinsic quality referred to as the network organization as a whole from the other side [17] (**Figure 3**).

According to the vision of this model, studies on the behavior of systems from the perspective of greater efficiency of performances consider that, when environmental conditions highlight the limits of a greatly integrated and autocratic organization, the drive toward decentralization begins, finding application in the fragmentation of the production process. Its extreme application uses point analysis and a focus on quality because of single actions [18, 19].

Structure typology	Integrated and Autocratic Network	Decentrated Network	Participatory Network	Centralized Network
Macro structure			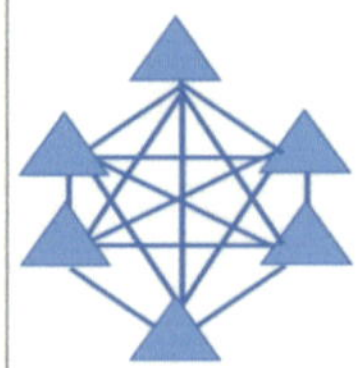	
Main features	Large dimensions with internal functions integrated	Large, middle and small dimensions with specializations induced by externalizations of leader organizations	Middle and small dimensions organized by network with no prevailing leader	Prevailing leader which distribute specialization owing their strategies of development

Figure 2.
Basic structures of network governance/dimensions.

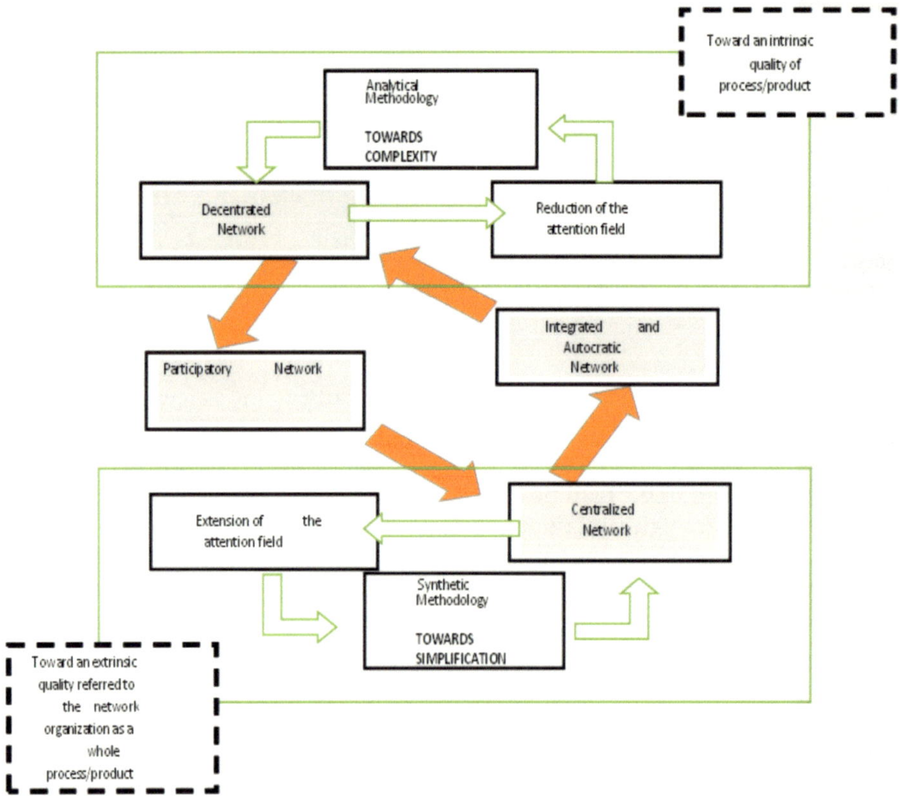

Figure 3.
Evolutionary cycle of network typologies.

This process is achieved by reducing the field of attention of analytical theories and methodologies that constitute a single package of phenomena that leads to a decentralized network as part of the simplification cycle [20].

With the rationalization of production and its greater repeatability, this process eventually facilitates the transition to the outsourcing of product components in external business units that become increasingly complex. External units result in greater autonomy from the parent company, favoring decentralization. Consequently, the evolutionary process moves toward a participatory network organization with specializations related to outsourcing relations centered on the production supply [21].

This evolution led some of these companies to switch to the production of increasingly complex components, allowing some of them to assume pre-eminent dimensions using positions of competitive advantage. In this phase the attention is focused on the extrinsic quality of the production, referring to the company's overall organization and the local environment from which it draws its culture. This process highlights the cycle of complexification with an expansion of the field of attention, applying synthetic methodologies, such as field analysis, and triggering a process that will again lead to centralized organizations [22].

The idea is to apply this interpretative model to the network, in general, considering that the context of a network is not strictly referred to a business organization but can be extended to a family, a condominium, a neighborhood, a city, a state, or the whole world. Once the network elements have been referred to as coherent, understanding models, the analysis can focus on the links between elements [3].

3. The order of ties

The distinction between direct (or first-order) and indirect ties considers two possible structures of a simple network (**Figure 4**) related to the question of governance and performance efficacy. In the network in **Figure 4b**, which contains the same number of elements as **Figure 4a**, indirect ties are limited to the second order (second-order links). However, in a complex network, the number of indirect ties could increase to a hyperbolic order.

Research conducted by the author evidenced that, in networks, the number of first-order ties changes according to the increase of components in a linear way and can be related to their intensity, represented in business organizations by the company turnover (**Figure 5**) [23].

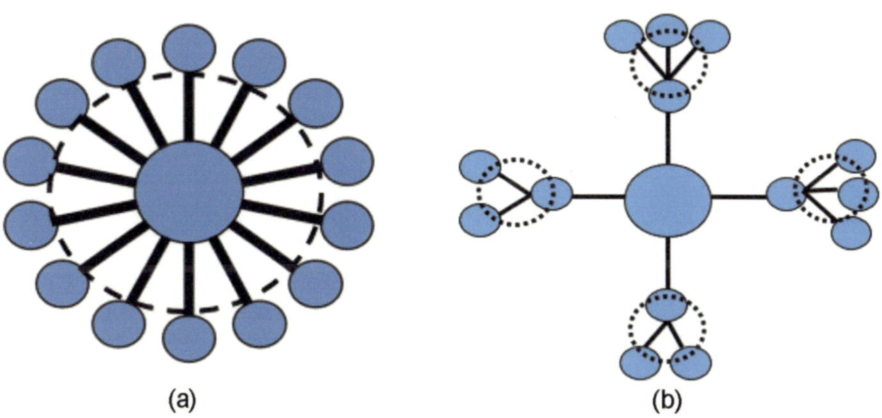

Figure 4.
First- and second-order ties (dashed lines): (a) first-order ties; (b) second-order ties.

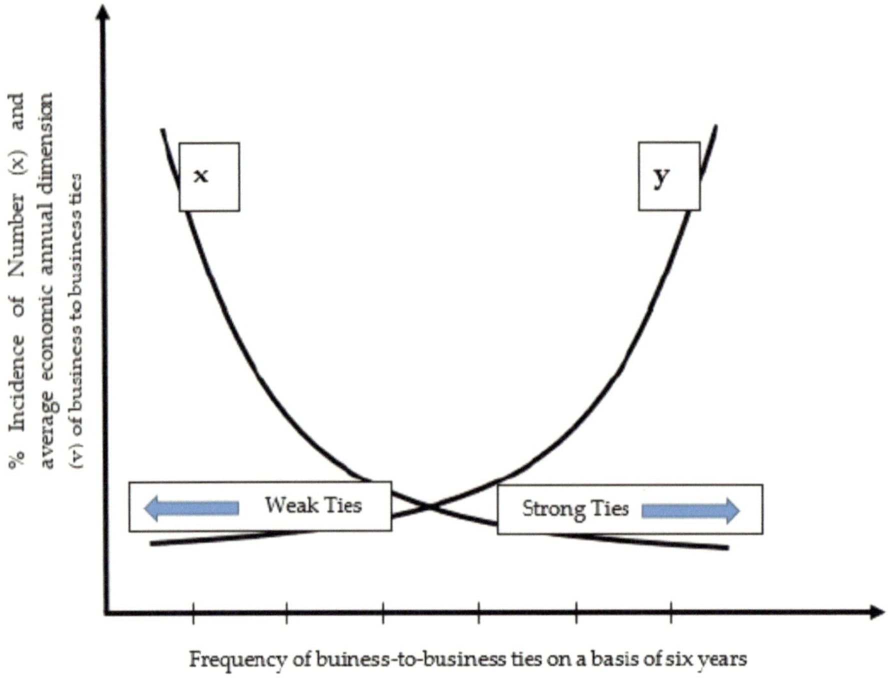

Figure 5.
Curves of numerousness (x and y) and 6-year frequency of business-to-business ties.

The two curves reveal that most of the intensity of ties, represented in business organizations by the turnover company (*y* curve), is concentrated on a limited number of business-to-business (B2B) relationships.

In terms of the network, the concrete dynamics of relationships derived from human behavior demonstrate that most of the ties are related to desultory relationships, whereas only a restricted number of links are connected to the highest intensity of relationships [24]. These trends, empirically detected in a sample of Italian small and medium enterprises (SMEs), were confirmed by other researches on supply chains and network management [25].

Particularly, the trend of B2B is recognized as strictly related to the customer and supplier selection process. This novelty, together the inadequacies of the current studies on the topic, orient the future direction of research on supply networks [26, 27].

Similar trends were detected on links connected to social networks in which Granovetter introduced the concept of strong and weak ties with particular attention to economic and organizational links and, in a wider sense, to the social system [28]. Weak ties in social systems are defined as most influential links not reinforced by mutual friendships. In contrast, strong ties are supported by direct, emotional links derived from the custom of relationships.

In the curves, in **Figure 5**, regarding the frequency of B2B ties, the distinction between the two categories is illustrated. Furthermore, the extension of the examined strength of the links to the structures of the technical networks is reasonable when they are applied to a great range of individuals, organizations, and urban structures [29].

Accordingly, in the research on their behavior, the analysis of link strength is applied with particular attention regarding the analysis of the link intensity according to the distribution of first-order ties [30].

4. The second-order ties

Second-order ties imply the existence of a hierarchical position from which the link is connected in an exclusive derivation or sharing with other elements directly connected to junction or branch points. This structural condition, displayed in **Figure 3b**, leads to the principle of "Graicunas Span of Control" (GSOC) [31] in the organization of the network, which Graicunas shared with Lyndall Urwick, one of his best estimators and master of scientific management in 30 years [32]. This principle was also listed in the "Ten Principles of Management" recently reconsidered [33]. The GSOC principle defines seven as the number of direct subordinates a manager can adequately manage.

In organizational structures, this problem led to the assumption of a hierarchical-functional order and authority managed with intermediate positions whose number was related to the organizational dimensions and whose order depended on the dimensions of the organizational network and on the managerial structure [34].

This leads to the consideration that, while the number of network connections, which a position oversees, can theoretically be infinite, in practice the organizational performance control capabilities are limited.

Figure 6 shows the hyperbolic growth of second-order ties correspondent to the first ones as calculated (**Table 1**) by Graicuna's Equation [Eq. (1)] in which R is the number of second-order ties and n is the correspondent number of first-order ties.

$$R = n\left[\frac{2n}{2} + (n-1)\right] \tag{1}$$

After its codification in organizational principles, GSOC was submitted to several critics and cited by Simon [35] as an example of an inconsistent criterion, at best considered a proverb. Particularly, Simon underlined that the GSOC has a relative appliance, as it is adequate only to restricted situations in which other elements, such as technology, information systems, and the managerial process, allow its appliance [36].

Considering the connection between March and Simon, we note that in the Handbook of Organizations edited by March [37], the principle was mentioned and significantly quoted by Urwick [38, 39] for any case limited to closed systems [40].

The last topic in the evolution of networks brings the Graicunas principle back up to date with the limits of control systems in the face of the growing complexity of organizations and with an extension to processes of its interpretative potential.

5. GSOC and control processes

GSOC was examined concerning the measuring of hierarchical dimensions [41] in connection with the order of the hierarchy. Other authors observed that the GSOC had a relative relevance in connection with the simplicity of jobs, increasing its measurement while decreasing with the surge of complexity [42] and speculated on its connection with the spans of accountability, influence, and support. These paradoxes can also occur in simpler scenarios where the invasive network control process coexists with the continuous looming risk of collapse of the control systems [43].

Figure 6.
Trend of first- and second-order ties according to the Graicunas Eq. (1).

First order Ties	Second order Ties
1	1
2	6
3	18
4	44
5	100
6	222

Table 1.
Trend of 2 ns order ties according to Graicuna's Eq. (1).

The extension of the GSOC principle to procedures can be proposed by applying a different approach in which the communication process between different organizational positions is intended as a process comprising many steps (**Figure 7**). The number of these steps is related to the efficacy and efficiency of the control [44].

In the networks, the existence of many points or steps of control may be interactive, involving many steps and operations (**Figure 7b**) [45]. Consequently, the information processed in the network is subjected to the loss of components and the acquisition of others in continuous renewal (**Figure 8**), as in computational transfer [46]. This process includes the risk of uncorrected information and inaccurate control feedback.

Progress in supply chain management implemented innovations in the tracking system of goods and services, producing advancements in integrating operators and software. This progress prompts the question: How many steps could a manager control? How many steps can be detected effectively by an organizational position [47]?

Once more, the attention of scholars and practitioners was focused on the study of hyperlinks [48], in which the multiphase process links two elements from a

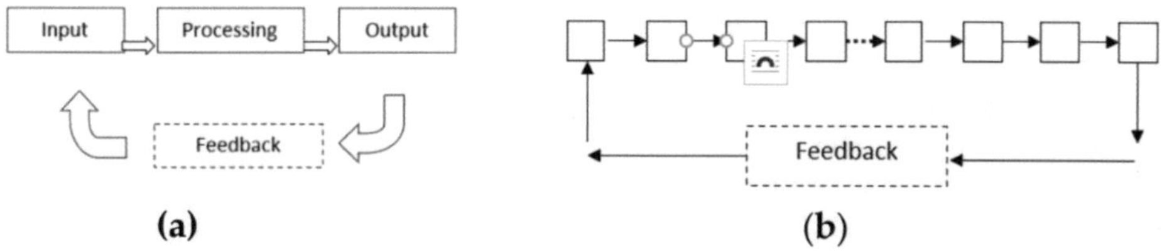

Figure 7.
(a) Basic process of control compared to the (b) multistep process typical of complex networks.

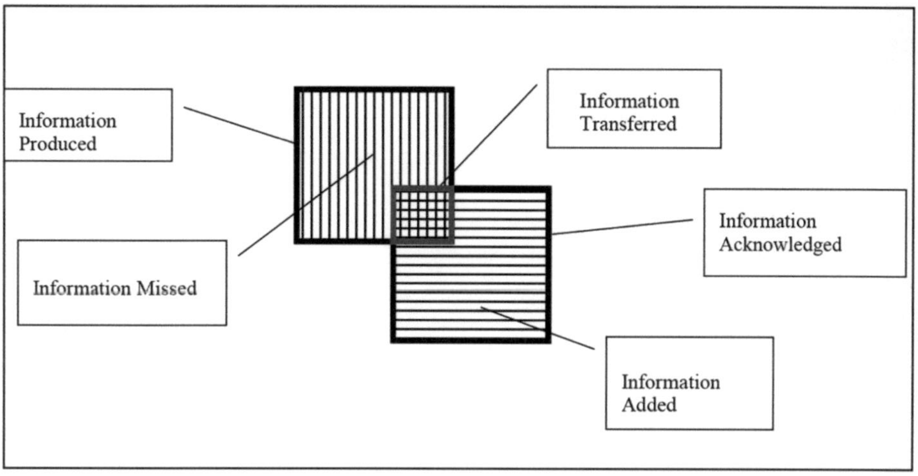

Figure 8.
Process of information renewal.

theoretical viewpoint, while a situation arises in practice in which the two extremes are isolated from each other when steps exceed some number [49].

This reasoning leads to the question, concerning the organizational boundaries of identifiable networks, considering the extension of the relationships between the component elements, and their strength and mutual influence [1]. Consistent with this, the extension of a network that comprises interactive elements can be conventionally established based on the boundary between weak connections subject to maximum renewal and strong connections featuring low renewal. This boundary can be identified while considering the bond intensity at the point of intersection of the x and y curves in **Figure 5**.

This led to the evolutionary model of organizational networks in which the unrestrainable renewal of ties produces a complexification of controls of each specific structure and consequently a structural evolution of organizations [50]. When limits of adaptation were overcome the evolution proceeds by a jump to another kind of structure in a cycle in which the contingent typology of the network is only a step.

6. Conclusions

With these considerations on network analysis, from a transdisciplinary perspective, the meaning of many measurements and indexing aimed at improving organizational performances invites us to consider the limits of the analysis and the complexity of the tools used for this purpose. A limit of the extension of the number of indices focusing on a few shared indices is conceivable in the future of network analysis [51]. There is also a need to assess the simplification of the analysis to originate systems that can be considered, not just on paper, susceptible to concrete and feasible control. Finally, one should consider the limitation of the boundaries of the network, which do not ensure adequate reliability of their consistency if certain thresholds of the degree of interdependence are exceeded, especially concerning control systems.

Funding

This research received no external funding.

Conflicts of interest

The authors declare no conflict of interest.

Author details

Massimo Bianchi
University of Bologna, Bologna, Italy

*Address all correspondence to: massimo.bianchi@unibo.it

References

[1] Scott J, Carrington PJ. SAGE Handbook of Social Network Analysis, SAGE 2011 Introduction, Management Communication Quarterly. London: SAGE Journals; ISBN 978-1-84787-395-8

[2] Gethmann CF, Carrier M, Hanekamp G, Kaiser M, Kamp G, Lingner S, et al. Interdisciplinary Research and Trans-Disciplinary Validity Claims. Vol. XVI. Springer Cham; 2015. p. 195. DOI: 10.1007/978-3-319-11400-2

[3] Aldrich H, Wetten DA. Organization-sets, action sets and networks: Making the most of simplicity. In: Nyström PC, Starbuck, editors. Handbook of Organization Design. Oxford University Press; 1981. pp. 385-408. ISBN: 9780198272410

[4] Varela LM, Rotundo G, Auloos M, Carrete J. Complex Network Analysis in Socioeconomic Models. Physics and Society, Springer; 2015. pp. 209-245. DOI: 10.48550/arXiv.1403.6767

[5] Chen J, Fang S, Ishii H. Fundamental limitations and intrinsic limits of feedback: An overview in an information age. Annual Reviews in Control. 2019;**27**:155-157. DOI: 10.1016/j.arcontrol.2019.03.011

[6] Mella P, Pellicelli M. How myopia archetypes lead to non-sustainability. Sustainability. 2018;**10**(1):21. DOI: 10.3390/su10010021

[7] Hu TY, Mahmassani H. Day-to-day evolution of network flows under real-time information and reactive signal control. Transportation Research Part C: Emerging Technologies. 1997;**5**(1):51-69. DOI: 10.1016/S0968-090X(96)00026-5

[8] Jahn T, Bergmann M, Keil F. Transdisciplinarity: Between mainstreaming and marginalization. Ecological Economics. 2012;**79**:1-10

[9] David Knoke D, Song Yang S. Social Network Analysis, 3rd Edition, Sage, 2020

[10] Caldarelli G. A perspective on complexity and networks science. Journal of Physics: Complexity. 2020;**1**(2). DOI: 10.1088/2632-072X/ab9a24

[11] Greiner LE. Evolution and revolution as organizations grow. In: Asch D, Bowman C, editors. Readings in Strategic Management. Palgrave; 1989. pp. 373-387. DOI: 10.1007/978-1-349-20317-8_25

[12] Kuhn T. The Structure of Scientific Revolutions. 3rd ed. University of Chicago Press; 1996. p. 210. DOI: 10.1007/978-1-349-20317-8_25

[13] Popper K. The Logic of Scientific Discovery. Routlegde; 2002. p. 544. DOI: 10.4324/9780203994627

[14] Solé RV, Valverde S. Information theory of complex networks: On evolution and architectural constraints. In: Ben-Naim E, Frauenfelder H, Toroczkai Z, editors. Complex Networks. Lecture Notes in Physics. Vol. 650. Berlin, Heidelberg: Springer; 2004. DOI: 10.1007/978-3-540-44485-5_9

[15] Serrat O. Social network analysis. In: Knowledge Solutions. Singapore: Springer; 2017. DOI: 10.1007/978-981-10-0983-9_9

[16] Alvargonzález D. Interdisciplinarity, transdisciplinarity, and the sciences. International Studies in the Philosophy of Science, Taylor & Francis. 2011;**25**(4):387-403. DOI: 10.1080/02698595.2011.623366

[17] Morselli GD, Nowick K. Construction, comparison and evolution of networks in life sciences and other disciplines. Journal of the Royal Society Interface. 2020;**17**(166). DOI: 10.1098/rsif.2019.0610

[18] Whitbred R, Fonti F, Steglich C, Contractor N. From microactions to macrostructure and back: A structurational approach to the evolution of organizational networks. Human Communication Research. 2011;**37**(3):404-433. DOI: 10.1111/j.1468-2958.2011.01404.x

[19] Pilny A, Proulx DJ, Dinh L, Bryan AL. An adapted Structurational framework for the emergence of communication networks. Communication Studies. 2017;**68**(1):72-94. DOI: 10.1080/10510974.2016.1262886

[20] Reijers HA. Business process management: The evolution of a discipline. Computers in Industry. 2021;**126**(1). DOI: 10.1016/j.compind.2021.103404

[21] Ying Liu Y, Aron R. Organizational control, incentive contracts and knowledge transfer in offshore business process outsourcing. Information Systems Research. 2015;**26**(1):81-99. DOI: 10.1287/isre.2014.0550

[22] Weidlich W. From fast to slow processes in the evolution of urban and regional settlement structures. Discrete Dynamics in Nature and Society. 1999;**3**(1):137-147. DOI: 10.1155/S1026022699000175

[23] Bianchi M, Paganelli G. The renewal of business ties and the reproduction of networks with customers and suppliers. Economia Aziendale Online International Quarterly Review. 2017;**8**(4). Available from: www.ea2000.it. Electronic ISSN 2023-5498

[24] Bianchi M. Trust impact in the renewal of management systems: Theoretical problems and empirical evidences for boundary maintaining. Economic Research. 2004;**17**(2):21-30. DOI: 10.2139/ssrn.1090909

[25] Bianchi M. Inter-company relationships in the survey and assessment of organizational. In: Proceedings of the XXIII Convegno Italian Academy of Business Economy. McGraw-Hill; 2000. pp. 211-212

[26] Ounnar F, Pujo P, Mekaouche L, Giambiasi N. Customers-Suppliers Relationship Management in an Intelligent Supply Chain. Production Planning and Control. 2007;**18**(5):377-387. DOI: 10.1080/09537280701403736

[27] Fengru C, Guitang L. Global Value Chains and Production Networks: Case Studies of Siemens and Huawei. Academic Press; 2019. p. 216. DOI: 10.1016/C2017-0-01798-6

[28] Granovetter M. Economic action and social structure: The problem of embeddedness. American Journal of Sociology. 1985;**91**(19):481-210. DOI: 10.1086/228311

[29] Granovetter M. Getting a Job: A Study of Contacts and Careers. The University of Chicago Press. 2017. pp. 73-77. ISBN 022651840X,9780226518404

[30] Sweldberg R, Granovetter M. The Sociology of Economic Life. 3rd ed. London: Routledge; 2018. p. 594. DOI: 10.4324/9780429494338

[31] Graicunas V. Relationship in organization. In: Urwick L, Gulick L, editors. Papers on the Science of Administration. Institute of Public Administration; 1937

[32] Urwick LVA. Graicunas and the span of control. Academy of Management

Journal. 1974;**17**(2):349-354. DOI: 10.5465/254987

[33] Hammond TH. In defence of Luther Gulick's "notes on the theory of organization". Public Administration. 2007;**68**(2):143-173. DOI: 10.1111/j.1467-9299.1990.tb00752.x

[34] Meier KJ, Bohte J. Span of control and public organizations: Implementing Luther Gulick's research design. Public Administration Review. 2003;**63**(1):61-70. DOI: 10.1111/1540-6210.00264

[35] Simon HA. Administrative Behavior. A Study of Decision-Making Processesein Administrative Organization. 3rd ed. The Free Press, Collier Macmillan Publishers; 1976. p. 414. DOI: 10.4236/aasoci.2020.104009

[36] Klar D, Kotowsky K, editors. Complex Information Processing. The impact of Herbert Simon Psychology Press; 2013. p. 480. ISBN 1134739419, 781134739417

[37] March JG, editor. Handbook of Organizations. Routledge; 1965. p. 1268. ISBN 9781138975750

[38] Urwick LF. The span of control-some facts about the fables. Advanced Management. 1956;**21**(1):5-15. Digitalized from IA1533314-01

[39] Urwick LF. Why the so-called "classicists" endure. Management International Review. 1971;**11**(1):3-14. Available from: https://www.jstor.org/stable/40226548

[40] Bennis WG. Review of the social psychology of organizations., by D. Katz R. L. Kahn. American Sociological Review. 1966;**31**(5):745-745. DOI: 10.2307/20918

[41] Mihm J, Loch CH, Wilkinson D, Huberman BA. Hierarchical structure and search in complex organizations. Management Science. 2010;**56**(5):4-903. DOI: 10.1287/mnsc.1100.1148

[42] Simons R. Designing high-performance jobs. Harvard Business Review. 2005;**83**(7/8):54-62. PMID: 16028816

[43] Zhang XM, Han QL, Ge X, Ding D, Ding L, Yue D, et al. Networked control systems: A survey of trends and techniques. IEEE/CAA Journal of Automatica Sinica. 2020;7(1):1-17. DOI: 10.1109/JAS.2019.1911651

[44] Bianchi M, Tampieri L. Graicunas principle and hyperlinks in the prevention of audit failures: A challenge for auditors and ICT audit supports. International Journal of Auditing Technology. 2013;**1**(3/4):1757-8752

[45] Ge X, Yang F, Han Q. Distributed networked control systems: A brief overview. Information Sciences. 2017;**380**:117-131. DOI: 10.1016/j.ins.2015.07.047

[46] Nakagawa T, Naruse K, Maeji S. Optimal checking times of backup operation for a database system. In: Tomar GS, Chang RS, Gervasi O, Kim TH, Bandyopadhyay SK, editors. Proceedings of the advanced Computer Science and Information Technology: Second International Conference. Springer; 2010. pp. 136-138. ISBN: 978-3-642-13346-6

[47] Hayer MR, Oluyisola OF, Standhagen JO, Semini MG. Exploring the challenges with applying tracking and tracing technology in the dairy industry. IFAC PapersOnLine. 2019;**52**(13):1727-1732. DOI: 10.1016/j.ifacol.2019.11.450

[48] Alexander H. The hyperlink as organizing principle. In: Turow J,

Tsui L, editors. The Hyperlinked Society. University of Michigan Press; 2008. pp. 39-55. ISBN: 978-0-472-07043-5

[49] Siegenfeld AF, Bar-Yam Y. An Introduction to Complex Systems Science and its Applications. Complexity. Wiley; 2020. p. 16. DOI: 10.1155/2020/6105872

[50] Abatecola G. Research in organizational evolution. What comes next? European Management Journal. 2014;**32**(3):434-443. DOI: 10.1016/j.emj.2013.07.008

[51] Cameron K, Whetten D. Developing Management Skills. Pearson; 2016. p. 720. ISBN-13:9780133127478

Section 2

Information Systems and Their Applications

Chapter 5

Secured Storage Mechanism for the Cloud-Based Banking Applications Using ECC

Gopinath V.

Abstract

The security of data becomes more vulnerable despite the various safety measures. The significant contribution of the chapter is to describe the design of a secured Cloud storage mechanism using ECC. Many banks are using cloud technology. However, it is not free from hacks and threats. Here comes the focus on Cloud Security and a lot of research work is happening around it. ECC-based secured cloud framework with a logging mechanism provides a high-level design of the SSL VPN and secured cloud mobile banking. It helps to secure data transfer for users. Software applications and web services are handled by maintaining the log files securely, and this security scheme shields the aligning of different kinds of data elements to each provider using an ECC algorithm. It is implementing a two-layered security system, which includes both private and public cloud storage environments, also providing a customized secure logging mechanism that will be encrypted by ECC. The TRA (banking) customers will get lots of benefits from this cloud storage mechanism. The proposed design acts like a buffer for end users, Internet service providers, and Banks, so it is more effective and secure in the Cloud environment.

Keywords: ECC, SSL VPN, XML, web services, cloud computing, logging, cloud computing, banking, security, transaction rich applications (TRA)

1. Introduction

In this modern world, secured communication is a significant need to perform high-end data transfers. In banking, the secure cloud storage mechanism is gaining importance and is yet to become omnipresent. The concern with the banking storage mechanism is that it requires immediate attention, such as rewards, precautions, confidentiality, perspective, usability, data management, unpredictable growth in the volume of transactions, and price reduction. Cloud computing provides everything as a Service (XaaS) to the banking sector in the form of Software-as-Service (LaaS) Which SaaS supports customer relationship management, accounting, invoicing, and Enterprise resource planning. PaaS is a suitable platform for applications. It helps in reducing the cost of IT Infrastructure & the spending on hardware devices & software applications can be cut down significantly. With LaaS businesses can procure those assets as a completely redistributed service. The same approach is applied for secured usage of a cloud storage mechanism with XML Web services and a logging mechanism to provide a high-level design of the SSL VPN to track the user activity and identify in case of any vulnerable security issues. In addition, it is implemented for a safe storage mechanism using ECC in transaction-rich applications (TRA). Many TRA makes use of storage mechanism 4 due to the service-oriented architecture, seamless accessibility, and other advantages of this advent technology. At the same time, it is vulnerable to hacks and threats. Hence securing this environment is of utmost importance, and many research works are being reported focusing on it. Cloud service provider gives base security, but it is not sufficient to handle the financial data in the cloud environment. Both the banking industry and the customer look for more protection when it comes to funding transactions like payroll, CRM, accounting, invoicing, Enterprise resource planning, etc. Security Breaches are not tolerable in the banking sector. In the current environment of cloud computing. TRA has got only a single layer of security. So, the mechanism to provide high-level security with cost-saving, high performance, and bandwidth is needed. The proposed system is aimed at providing a high-level design of the SSLVPN with ECC, which is applied to a private cloud with a secure logging mechanism. The model has experimented with java coding and open VPN with the MS Azure cloud system. SOAP UI tool has been used for performance analysis, which is deployed into the cloud web server [1, 2].

The design applied to a private cloud environment, in addition to providing a customized secure logging mechanism that will be encrypted by ECC, ensuring data security and privacy protection of entire user activities in the cloud. To launch the ECC-based SSL VPN, the private cloud VPN frameworks should be added to the system.

The same approach is applied in the cloud storage mechanism for the TRA (banking) customers can get lots of benefits [3, 4].

- Utilization of Time: Customers can use it 24*7 hours.
- Increase Adaptability: It helps banks to enjoy the promotion of adaptability ratios and operating leverage [5, 6].
- Decrease Invest Amount: The banks are not ready to invest a large amount to purchase software, hardware, and related workforce for the usage of cloud computing.
- Security Comparison: ECC-based secure cloud storage mechanism for TRA is more sheltered than online and internet banking.

It provides an additional layer of security with ECC. While the banking customer is connecting to the cloud, it connects the P2P network with a second layer of security SSL and ECC [7], and the banking application utilizes the same set of ECC digital keys.

2. Design of Secured Cloud Storage Mechanism

ECC-based secure cloud framework with logging mechanism is to provide a high-level design of the SSL VPN and secure cloud mobile banking. It helps secure data transfer for the users [8–10]. Software applications and web services are used with secure log files maintained. **Figure 1** shows the high-level design of the Cloud storage mechanism.

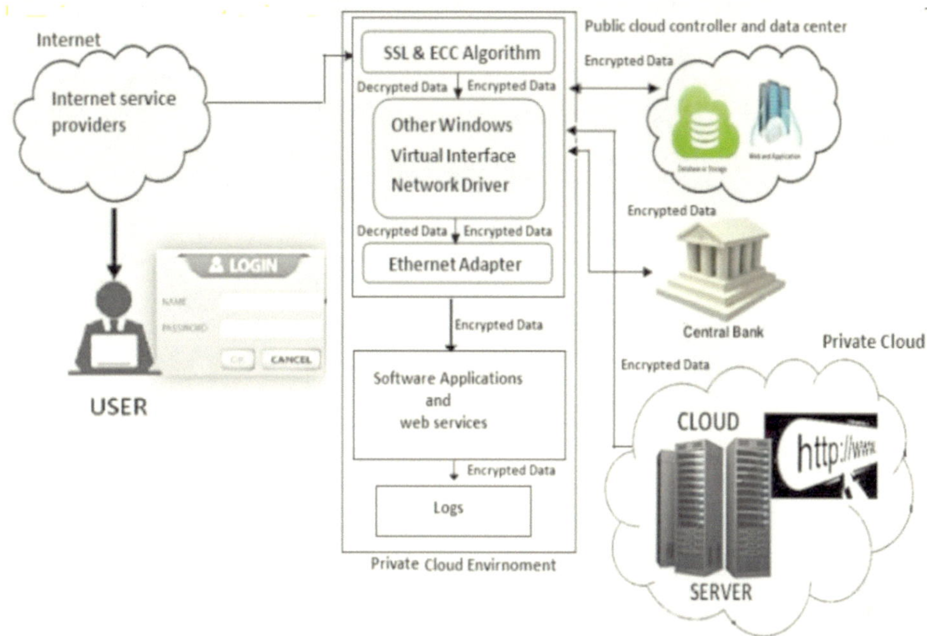

Figure 1.
High-level designs for cloud storage mechanism.

The design will provide a two-layer security system, which is included in the private cloud as well as a public cloud storage environment, along with a customized secure logging mechanism; the entire system will be encrypted by SSL-based ECC [11, 12]. Cloud storage mechanism for the TRA (banking) customers will get lots of benefits. The proposed design acts middle layer for the end user, Internet service provider, and the Bank.

3. Design of Secured Cloud Storage Mechanism

The architecture of a secure cloud TRA is explained in **Figure 2**. Movable equipment and the remote systems are connected to the mobile machinist and wireless access point. These are connected to either an entry point or satellite via the base station. The mobile user's request and response are delivered to cloud banking through the presently available SSL VPN with an additional layer of protection from ECC. **Figure 2** Architecture of secure TRA [1].

Central processors that are linked to the mobile network receive the mobile users' requests as well as send details such as user IDs and user location to banking database storage. The projected security theory enhances the level of protection that currently supports mobile banking and facilitates the realization of P2P networks.

The following are the objectives of the proposed framework:

- Assuring information security and privacy protection of entire mobile/remote user's activities in the cloud. ECC-based SSL VPN.

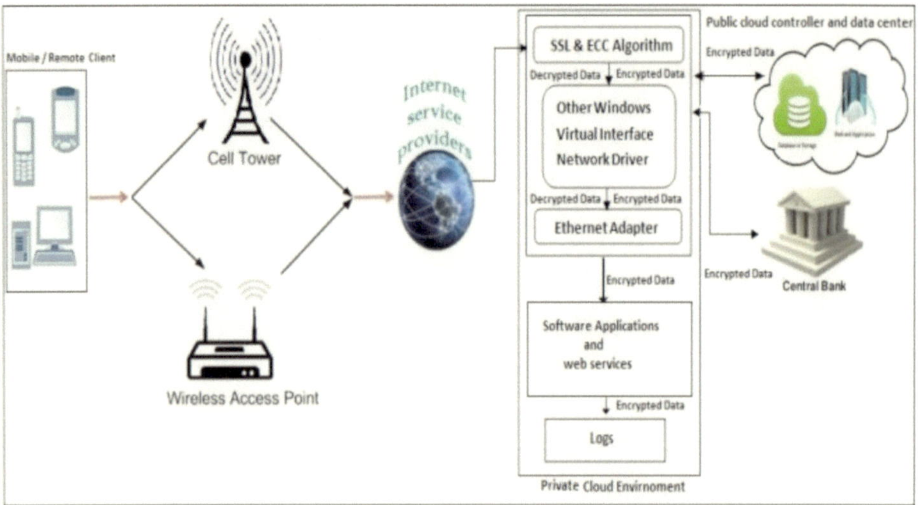

Figure 2.
Architecture of secure TRA.

- ECC encryption helps to provide less bandwidth, computing power, and memory for creating customized security for data packets.

3.1 Storage mechanism for transaction rich application

The cloud storage mechanisms are exclusively designed for cloud-based supply. Like how the physical server can release virtual server images, similarly, the device's instances can be made virtual. They can provide fixed-increment capacity allocation in support of the pay-per-use mechanism. Via cloud storage devices, stored data can be exposed with the help of remote access. It gives general logical elements of data storage, such as:

- The Group of data that are stored in the folders are called files.
- The lowest level of storage is called blocks.
- The collections of data are called datasets.
- Objects are called metadata.

There is a huge risk of data loss, it is essential to recover the data when there is an issue and failure. However, customers are ensured of data availability post such issues. So, it is essential to provide data security in the cloud. The proposed design provides a secure cloud storage mechanism for TRA.

Figure 3 [1] shows the data request of the user via the internet to access the private cloud data servers and applications to utilize the Bank's web server and a firewall. The design is trialed with java coding 43 and inbuilt ECC algorithm into the webserver. Logs are maintained in the DB tables as well as the web server, and the user accesses the web server system with the help of SOAP UI. The user hits directly the bank's web

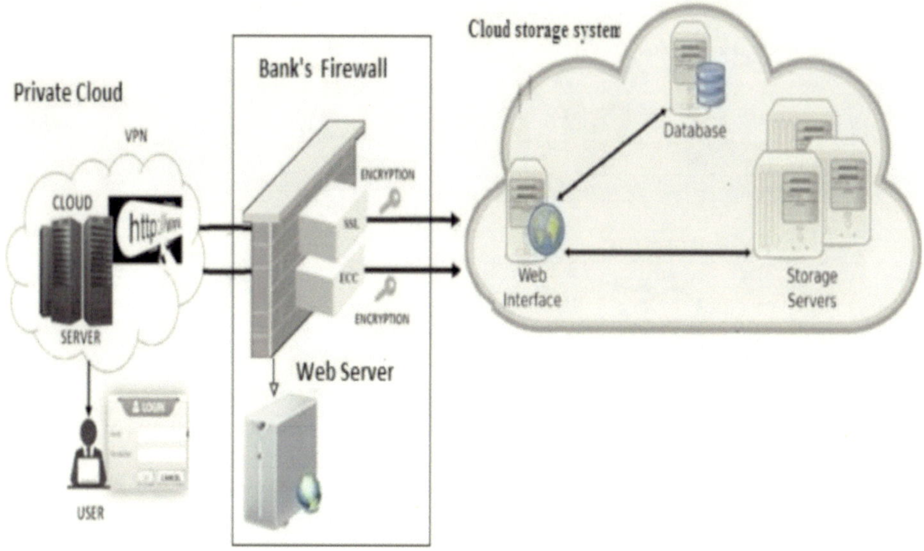

Figure 3.
Cloud storage mechanisms for TRA.

server and firewall, and data packets are sent as encrypted files for both requests and responses from the cloud storage system.

3.2 Cloud-based ATMs architecture using ECC

An ATM is used after installing the package, and it will be ready for user transactions. Usually, the ATM package consists of an operating system, machine third-party software and hardware packages, and core banking software; normally, this package deployment takes more than 4 hours to get completed. So, each machine installation takes 8 hours daily to be available to users. Banks are investing a huge amount in this activity. For increasing transactions, many banks are extending their activity by using ATMs. ATMs have become more user-friendly nowadays, which is a must need for the customer to use on a day-to-day basis; this promotes establishing banking. As there is a need for creating more ATMs to provide easy access to customers, banks need to invest more in setting up these machines, which can now be reduced by using the cloud. However, security becomes a concern worry about using this installation in a cloud environment. Here, a private cloud is used to install/deploy the package in a single trigger; within 4 hours entire installation is performed for more than 1000 ATMs; also, it provides the double-level security to connect the Host server and the core bank. Below is **Figure 4**. Shows the ATM package deployment with the help of a secure cloud environment.

Here, the package is deployed in the private cloud host server itself. Also, it is possible to increment the ATM node by more than 1000 in a single installation time, so this method is more cost effective and also provides high-level security for the customer data.

Figure 5 discusses the ATM package and machine architecture using a secure cloud environment. The core banking server interacts with the Cloud hosting server,

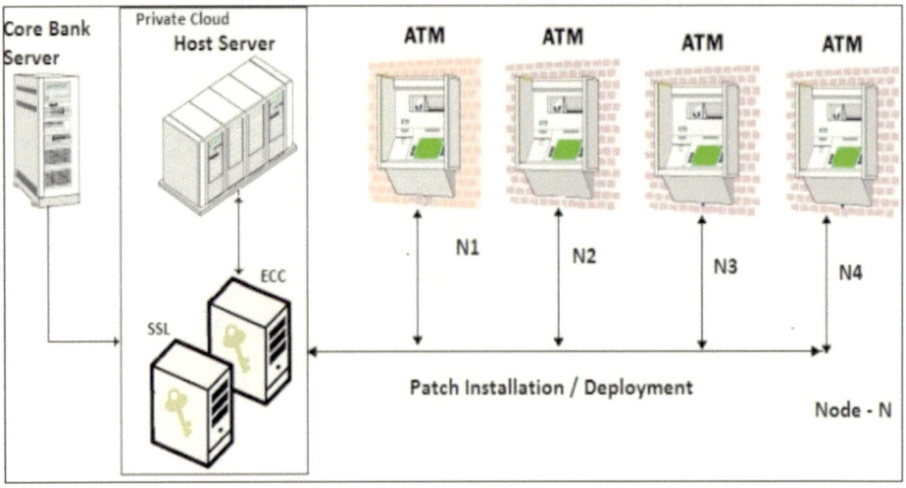

Figure 4.
ATM package deployments using secure cloud.

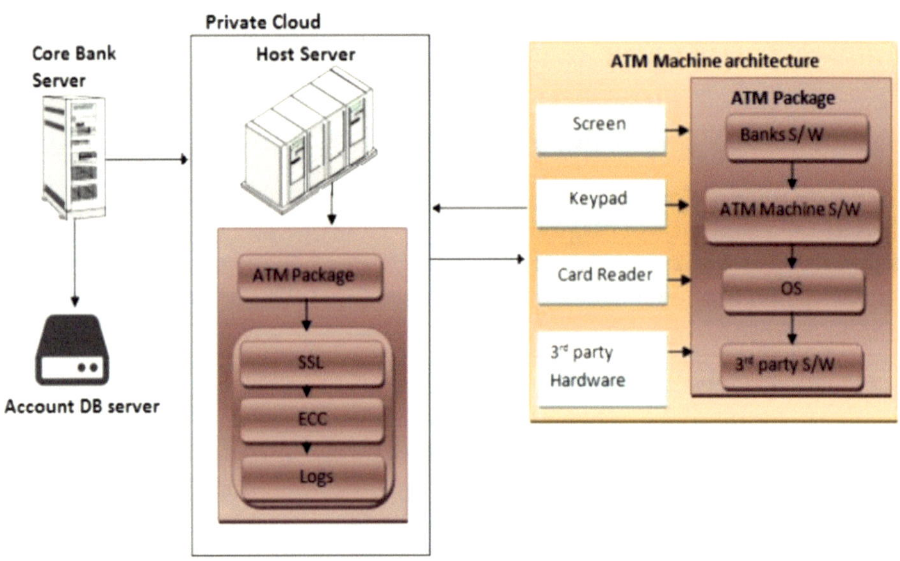

Figure 5.
ATM package and machine architecture using the secure cloud.

which has the ATM software packages, which are customized bank screens, ATM software, other 3rd party machine software, and operating system.

ATM packages have two-layer securities like SSL and ECC which gave high-level security of the bank's data and package while the installation got initiated.

The proposed design will give high standard security and more cost savings to the banks. Therefore, they can meet their customer demands.

4. Performance analysis for the cloud-based TRA

Cloud computing is characterized by accessibility at whichever instance is desired, with the reason for the reduction in costs and increase in operational efficiency. The ECC VPN that connects to cloud computing should possess the above-mentioned characteristics. The network atmosphere should be trouble-free to set up, modernize, or erase connections based on the demand for cloud computing, to ease the complex system management.

The below **Figure 6** shows the most vulnerable cyber-attack on the bank in the cloud environment; in this measurement, a high percentage of cybercriminals risk occurs in web applications and web services in the cloud environment. Many banks have identified and reported 89% of breaches in the last few years. Many a time, SQL injection and malware have accessed sensitive information in Cloud storage mechanisms like DB, web applications, and web services. It is hard to find web application attacks because banks have millions of users accessing their sites at the same point in time. So, spotting cybercriminals is a difficult task. The second highest-rated risk is the DoS attack; it completely 58 hacks down a system, and shuts down a machine/network, making it inaccessible to its intended users. The rest of the other attacks is very minimal to the overall bank attacks in the cloud enlivenment.

In **Table 1**, most of the efforts in the below-mentioned checklist talk about the number of vulnerabilities and threats focused on Cloud Computing. It describes the analysis report that is related to the analysis, and test used in cloud environments, and it specifies that the cloud service representations are open to these vulnerabilities and threats. More prominence on threats has been applied that are allied with statistics being stocked up and practiced remotely, sharing resources, and the usage of virtualization [3, 4].

This section presents the result and performance analysis for the proposed secure cloud storage mechanism for TRA. The design has experimented with java coding and an inbuilt ECC algorithm, which is deployed in a cloud environment. SOAP UI tool has been applied for performance analysis; it is deployed into the cloud web

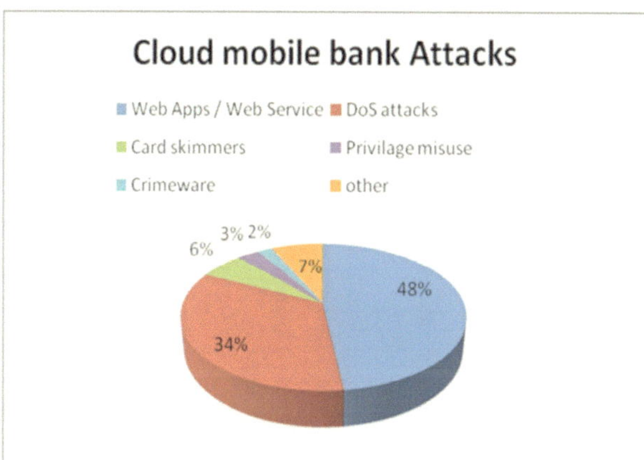

Figure 6.
Vulnerable analysis report to cyber-attack for the bank.

S. No	Vulnerability & threats analysis	Existing method (RSA)	Proposed method (ECC+ SSL)	Analysis report	Existing frameworks	Proposed framework
1	Secure even after the loss of user identity and password	✓	✓	Configuration Management	Single	Single
2	Indexing of data, Keyword Search	✓	✓	Fault Management	Simple	Simple
3	Insecure interfaces and APIs, Account or service hijacking	✓	✓	Performance Management	Integrated	Integrated
4	Unlimited allocation of resources, Denial of Service	✓	✓	Security	Support	support
5	Data-related vulnerabilities, Data leakage	✗	✓	Management	directory server	encrypted data and directory server
6	Vulnerabilities in Virtual Machines VM hopping	✗	✗	Accounting Management	Convenient	Convenient
7	Vulnerabilities in Virtual Machine Images, Data leakage	✗	✓	Confidentiality	File encryption not provided	Symmetric key
8	Vulnerabilities in Hypervisors, VM escape	✗	✓	Authentication	Only base-level security is provided.	Password-based advanced-level security is provided.
9	Vulnerabilities in Virtual Networks, VM escape, Customer-data manipulation	✗	✓	Access Control	Exposure to the normal area	Encryption of security area information
10	Log Security in Virtual Machine, Data Packets	✗	✓	Impersonation Attack	No impersonation	Two stages of user authentication

Table 1.
Vulnerability & Threats Analysis Report RSA vs. ECC.

server. With the help of this, the performance can be checked with load balance. Single usage of customized SOAP UI can trigger 1000 users at a time in the cloud storage server. Logs are stored in the database table as well as in the server and client system based on the load balance testing and performance tracking below **Table 2**. Shows the various comparisons of the proposed and existing systems in the Cloud storage mechanism [1].

Below **Table 3** ECC and RSA key size strength ratios are measured in the table; for example, the RSA system requires 7680 bits to provide security, whereas ECC requires only a 384-bit key to acquire the security and its key ratio stands at 1:20. So ECC provides greater efficiency in terms of key size and bandwidth; it means higher speed and lower power consumption [1].

Figure 7 discusses the key length comparison of the proposed ECC algorithm vs. the existing RSA algorithm [1].

The public key operation of ECC-160 is only 3.69 milliseconds, and it is 50% comparatively lesser than RSA-1024 and other keys. The flowchart given below explains the key generation time and required memory size for both ECC key and 1024-bit RSA key, 160-bit ECC is much better than RSA [13]. The protection measures for both 160-bit ECC key and 1024-bit RSA key are similar. Hence breaking

S. No	Cloud analysis report	Proposed system (ECC + SSL)	Existing system (RSA)
1	Upfront Investment	Low investment	High investment
2	Additional Hardware/IT costs	Not required	Yes, Required
3	All-Time costs	Predictable cost	Unpredictable cost (but maybe lower)
4	Degree of customizations	Less customizable in general	Greater ability to customize
5	Control of data security standards	Customers can control the data	banks can control the data
6	Confidentiality	Symmetric key	File encryption not provided
7	Authentication	Password-based advanced-level security is provided.	Only base-level security is provided.
8	Access Control	Encryption of security area information	Exposure to the normal area
9	Log Security in Virtual Machine, Data Packets	Provided and It is Secured	Not provided
10	Computational Overheads	Roughly 10 times that of RSA can be saved	More than ECC
11	Encryption Time	Much Faster than RSA	At good speed but slower than ECC
12	Decryption	Slower than RSA	Faster than ECC
13	Small Devices efficiency	Much more efficient	Less efficient than ECC

Table 2.
Comparison between existing and proposed framework.

Proposed ECC algorithm key size	Existing RSA algorithm key size	Key size ratio
224	2048	1:10
256	3072	1:12
384	7680	1:20
512	15,360	1:30

Table 3.
Key size strength ratio for the proposed system.

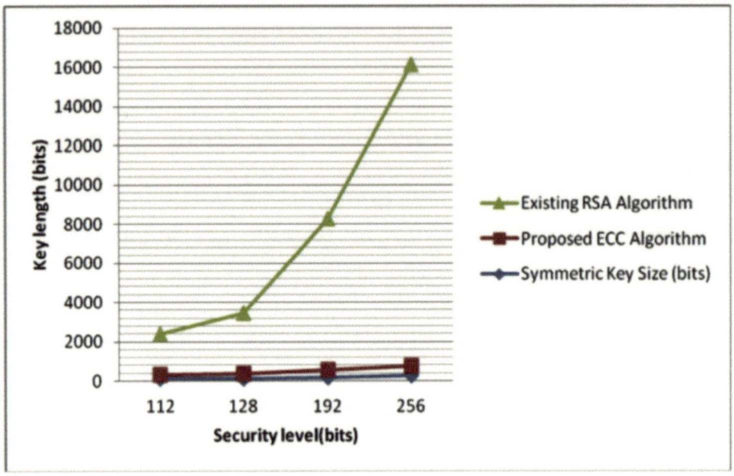

Figure 7.
Key length comparison for proposed and existing.

Algorithm	Key size	Key generation time (ms)	Required memory size (bytes)	Encryption/decryption time (ms)
Proposed Algorithm (ECC)	160	98	125	121
	224	177	140	183.1
Existing Algorithm (RSA)	1024	1312.7	313	388
	2048	6804.6	621	1867

Table 4.
Measure of proposed ECC vs existing RSA public keys.

a 160-bit key would be a hundred million times harder than breaking the 1024-bit key. In below **Table 4**, discuss the measure of proposed ECC versus existing RSA public keys.

In **Table 4**, the proposed public key is generated using the existing key algorithm, based on key generation time (ms), memory size (bytes), and encryption/decryption time. During the ECC key generation, the time taken to generate the RSA key and ECC key and size has a huge difference (ECC-224 and RSA-2048). Also, it is learned that during the signature generation process, ECC surpasses RSA. Conversely,

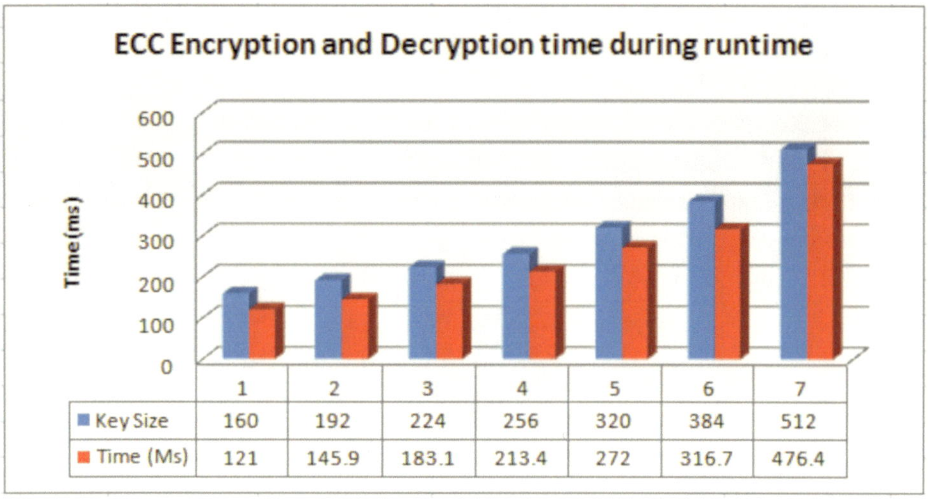

Figure 8.
Proposed algorithm encryption/decryption during runtime.

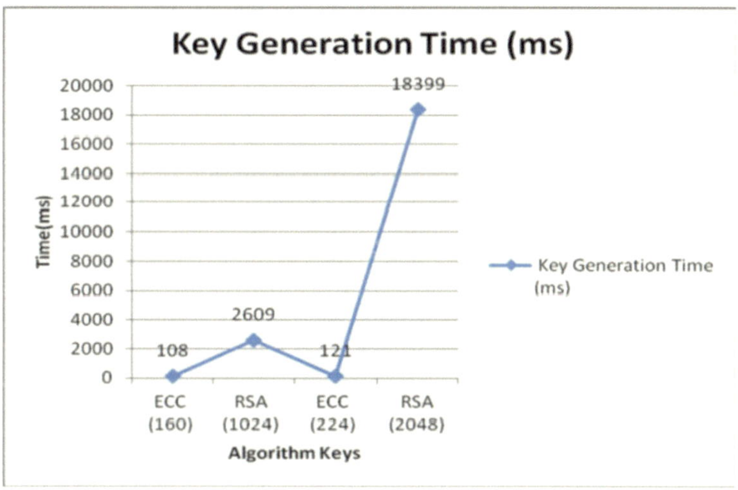

Figure 9.
Algorithm key generation time (ms).

RSA beats ECC in performance during the verification process and the Encryption/ Decryption process is ECC much better than RSA. The below **Figure 8** shows the encryption and decryption time during the run time of the proposed public key algorithm [13, 14].

The below **Figures 9–12** discusses the key size generation time, required memory size, and encryption/decryption time of the proposed ECC algorithm vs. the Existing RAS algorithm.

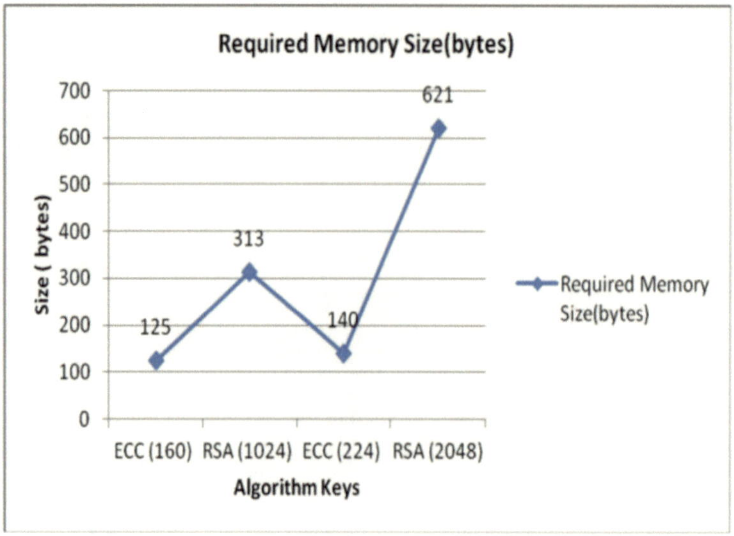

Figure 10.
Algorithm required memory size (bytes).

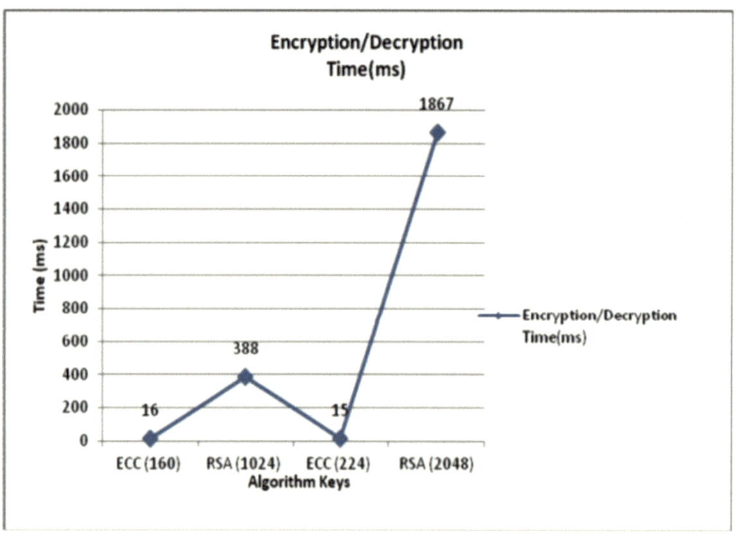

Figure 11.
Algorithm encryption/decryption time (ms).

The above experiment result tells us that the proposed ECC algorithm is much more efficient than the existing RSA algorithm in Speed, memory size, and security [15, 16].

The analysis of encryption/decryption execution time for both the proposed design and the existing system has been discussed in the below report, experimented with the sensitive banking data, and file transfer in the cloud environment is measured in regards to file size and execution time (mile second). Results tell us that the

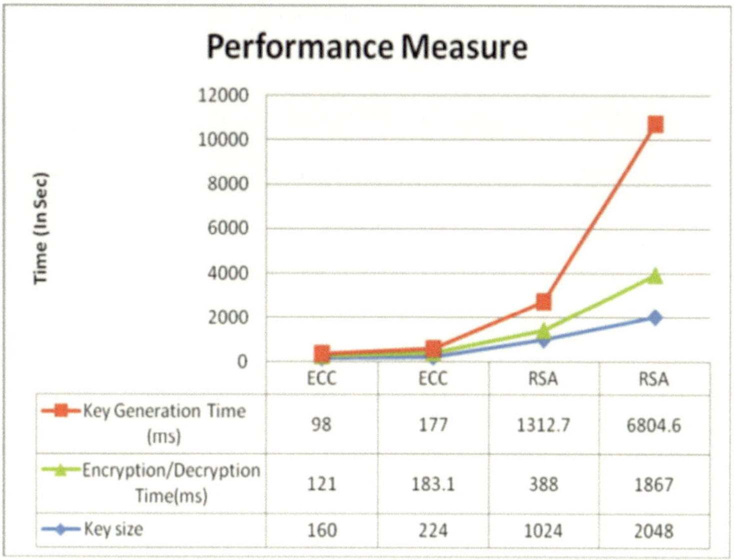

Figure 12.
Performance measure for ECC and RSA.

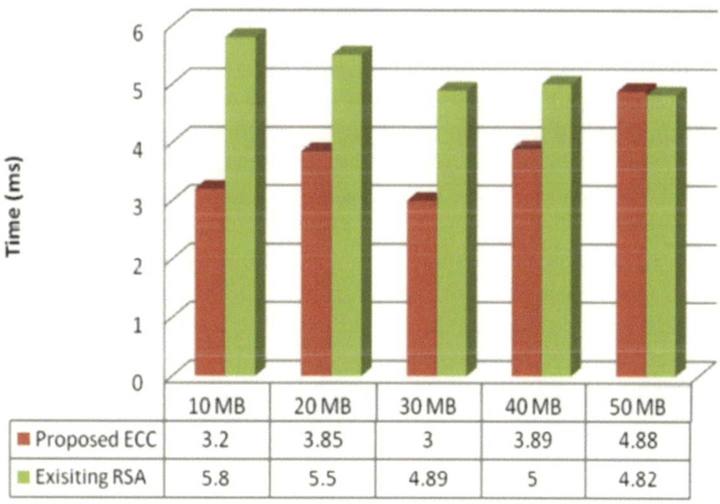

Figure 13.
Memory size analysis for the bank data on the cloud.

proposed design is better than the existing system, specifically bank data handle in the cloud environment.

Figure 13 discusses the memory size of the sensitive data transfer on the cloud environment, memory size, and the execution time of the encryption/decryption that is measured for the particular document. In execution, the proposed ECC is much better than the existing RSA; **Figure 14** discusses the experimental results clearly showing that the proposed security model can encode the data with very low processing time and a lesser amount of memory size in the cloud system.

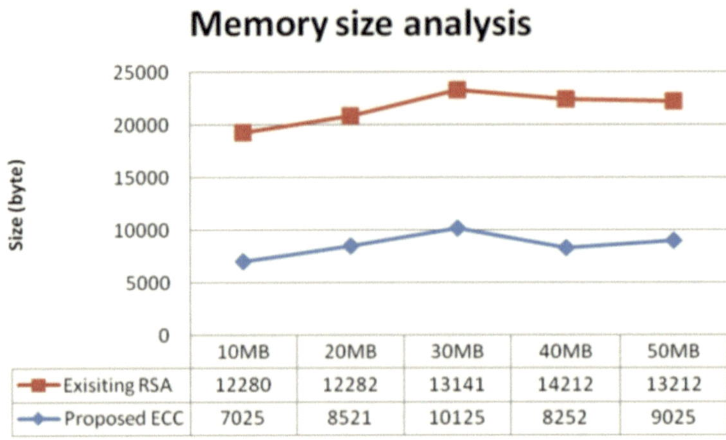

Figure 14.
Execution analysis for the bank data on the cloud.

5. Conclusion and future work

The proposed security scheme improves the level of protection currently supported in cloud storage mechanisms. The potential incompatibilities that arise from the simultaneous use of ECC, as well as the impact of user mobility on VPN operation, are considered, and detailed solutions are proposed. A design is proposed to focus on and improve the security level in the cloud environment. This research focuses on simulating and analyzing the performance of secure cloud storage. Though various methods are cited in the literature for the stabilization of cloud computing and portable device VPN connection [17–19], securities provided by these methods are not adequate for the protection of data packets. The proposed method provides two layers of security.

The work shall be extended to transaction-rich applications (Banking) as banking customers need more protection for fund transactions like payroll, CRM, accounting, invoicing and ERP, etc. Banks cannot afford the risk of a security breach. The current system has a single layer of security, so our proposed design will provide double-layer security with high performance and bandwidth.

Author details

Gopinath V.[1,2]

1 Sathyabama Institute of Science and Technology, Chennai, India

2 Tata Consultancy Services, Chennai, India

*Address all correspondence to: gopinathv0507@outlook.com

References

[1] Gopinath V, Bhuvaneswaran RS. Design of ECC based secured cloud storage mechanism for transaction rich applications. Computers, Materials & Continua. 2018;**57**(2):341-352

[2] Niyaz Ahamed N, Duraipandian N. Secured data storage using deduplication in cloud computing based on elliptic curve cryptography. Computers, Materials & Continua. 2022;**41**(1):83-94. DOI: 10.32604/csse.2022.020071

[3] Rajavarmana R, Vetriselvib T, Devic SS. Hybrid security system over banking transaction maintance by a Meta key. Turkish Journal of Computer and Mathematics Education. 2021;**12**(7):864-868

[4] Rehman S, Bajwa NT. Hybrid AES-ECC model for the security of data over cloud storage. Electronics. 2021;**10**(21):2673. DOI: 10.3390/electronics10212673

[5] Renju C. Pros and cons of mobile banking. IJSRP. 2014;**4**(10):1-5

[6] Agre C. Implementation of a cloud in the banking sector. International Journal of Computer Science and Information Technology. 2015;**3**(2):1168-1174

[7] Alemu M, Omer AM. Cloud computing security framework for the banking industry. HiLCoE Journal of Computer Science and Technology. 2014;**2**(1):79-85

[8] Vengala DVK, Kavitha D, Siva Kumar AP. Three-factor authentication system with modified ECC based secured data transfer: Untrusted cloud environment. Complex & Intelligent Systems. 2021. pp. 1-14

[9] Sridharan S, Arokiasamy A. Effective secure data storage in cloud by using ECC algorithm. Middle-East Journal of Scientific Research. 2017;**25**(1):117-127, ISSN 1990-9233. DOI: 10.5829/idosi.mejsr.2017.117.127

[10] Huang L, Zhang JZG, Zhang M. Certificateless public verification for data storage and sharing in the cloud. Chinese Journal of Electronics. 2022, 2020;**29**(4):639-647. DOI: 10.1049/cje.2020.05.007 Print ISSN 1022-4653, Online ISSN 2075-5597

[11] Chhabra A, Arora S. An elliptic curve cryptography based encryption scheme for securing the cloud against eavesdropping attacks. In: IEEE 3rd International Conference on Collaboration and Internet Computing (CIC). 2017. DOI: 10.1109/CIC.2017.00040

[12] Gopinath V, Bhuvaneswaran RS. Secure cloud computing with ECC web service and logging mechanism. Asian Journal of Information Technology. 2017;**16**(10):746-753

[13] Abbas SA. Improving data storage security in cloud computing using elliptic curve cryptography. IOSR-JCE. 2015;**17**(4):48-53

[14] Li Y, Zhang P. Security Analysis and Improvement of Elliptic Curve Digital Signature Scheme, Artificial Intelligence and Security. springer-ICAIS. 2019. pp. 609-617. SN - 978-3-030-24270-1July

[15] Gupta DS. A secure cloud storage using ECC-based homomorphic encryption. International Journal of Information Security and Privacy (IJISP). 2017;**11**(3):9. DOI: 10.4018/IJISP.2017070105

[16] Agrawal K, Gera A. Elliptic curve cryptography with hill cipher generation for secure text cryptosystem. International Journal of Computer Applications. 2014;**106**(1):18-24

[17] Khan IA, Qazi R. Data security in cloud computing using elliptic curve cryptography. IJCCN, ISSN: 2664-9519 (Online). 2019;**1**(1):46-52

[18] Gopinath V, Bhuvaneswaran RS. Design of Security System of portable device: Securing XML web service with ECC, Communication in Computer and Information Science. Vol. I. 2014. pp. 431-439

[19] Chawla BK, Gupta OP, Sawhney BK. A review on IPsec and SSL VPN. International Journal of Scientific & Engineering Research. 2014;**5**(11):21-24

Chapter 6

Impact of Digital Vehicle Identification Errors on Critical Information Systems

Roman Rak and Dagmar Kopencova

Abstract

In commercial, technical, administrative, record-keeping and forensic practice, unambiguous, unique identification of all objects with which we work is important. Vehicles are identified using a VIN (Vehicle Identification Number), which is a key identifier in everyday practice. This identifier is inserted into information systems in different ways. Manual copying from paper documents to record systems prevails. In practice, however, it turns out that this method has an average error rate of up to 8%. In recent years, the digital VIN, which is physically stored in the vehicle in its electronic control units, has started to be used for vehicle identification. The contribution deals with the description and evaluation of various ways of VIN entries in information systems, especially critical infrastructure, and analyzes their shortcomings and benefits. In the article, a thorough analysis of frequent errors in VIN is carried out and ways to eliminate them are suggested.

Keywords: vehicle emergency system, vehicle identification number, rescue, critical infrastructure, data quality, VIN decoding

1. Introduction

Data quality is an important attribute of every information system, for which we expect a certain functionality, efficiency, and reliability [1–3]. Data quality is crucial, especially for so-called key object identifiers (unique primary keys in computer terminology), which uniquely identify a given object [4]. In the area of motor vehicles and information systems, which store vehicle information for various purposes, the object identifier is the VIN (*Vehicle Identification Number*) [5–8] (**Figure 1**).

The quality and error-free entry of the VIN in information systems determines whether a vehicle, searched or examined for various reasons, will be found in the computer database at all [9, 10]. Today, the effectiveness of the police and other state security forces (including the fight against organized crime or terrorism) [11], the effectiveness of rescue services in the event of a vehicle accident (pan-European eCALL project [12], the control activities of state administration bodies, post-sales (service) and other services of the automotive industry, the services of insurance

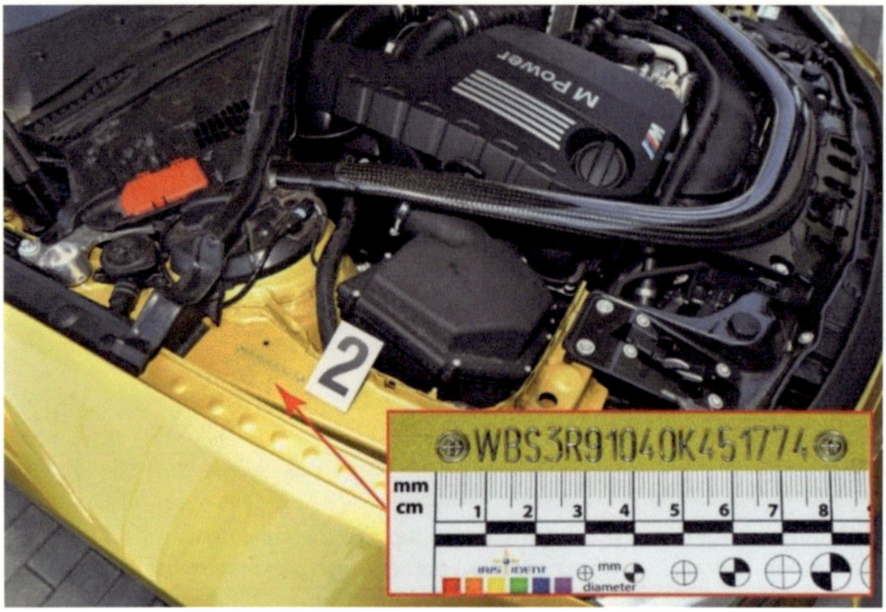

Figure 1.
Example of the location and appearance of a 17-digit VIN in a vehicle. Source: Roman Rak.

companies, leasing companies, and other various third parties in the commercial and noncommercial sector depend on the VIN quality.

The paper deals with the analysis of data quality in government information systems. However, this quality does not match the modern data collection, acquisition, and control technologies that these technologies offer today. There are numerous errors in the primary VIN identifier in the information systems because this identifier is still manually transferred from vehicle documents (where there may be errors, forged or altered) to both the state and private information systems without acknowledging that there may be a big number of errors [13, 14]. In specific cases, these errors can lead to fatal consequences—failure to find a stolen or safety defective vehicle, failure to provide the necessary information for the activities of the emergency services, i.e. in extreme cases, endangering the health and life of persons involved in a serious traffic accident [15, 16], frauds in car purchases, damage settlement, civil disputes, etc. [17].

2. Material and methods

The globally unique VIN identifier (*Vehicle Identification Number*) is defined worldwide using internationally valid ISO 3779:1983 standards (since 1986) – *Road vehicles -- Vehicle identification number (VIN) -- Content and structure*; ISO 3780:1983 – *Road vehicles -- World manufacturer identifier (WMI) code*, and ISO 4030:19831983 – *Road vehicles -- Vehicle identification number (VIN) -- Location and attachment.* The mentioned standards specify and implement the unambiguous vehicle identification worldwide.

A VIN is a string of alphanumeric characters of the precise length of 17 characters. To avoid visual similarity and inaccuracies, the O, Q, and I characters are prohibited. The VIN has three basic components (see **Figures 2** and **3**):

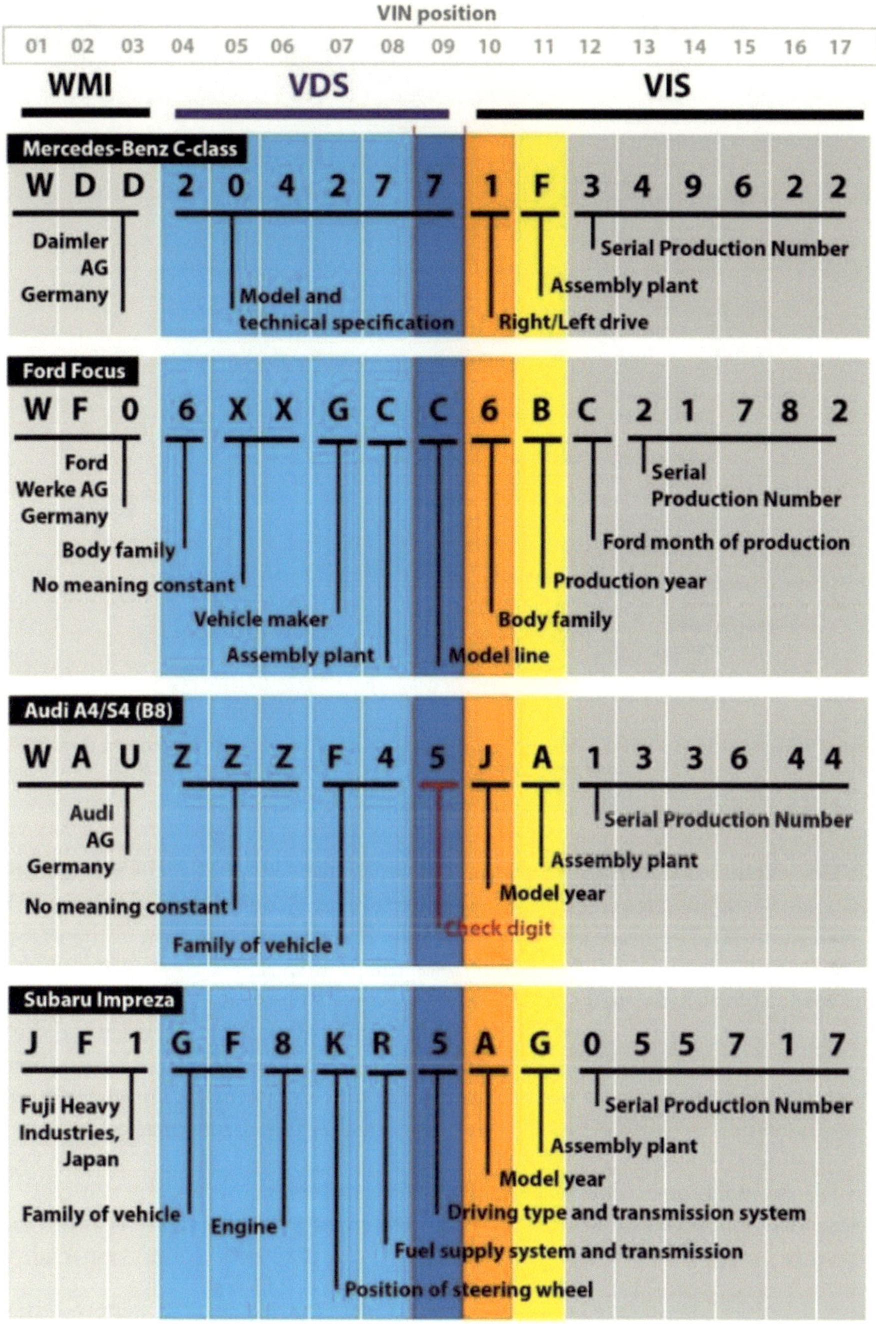

Figure 2.
Example of the diversity of VIN identifier structures used by different vehicle manufacturers. Some German or Italian manufacturers include meaningless strings like ZZZ, XX, or 000 in the VIN structure. Source: Roman Rak.

- **WMI—World Manufacturer Identifier.** Three-character sequences identify the vehicle manufacturer (factory make). This part is internationally standardized.

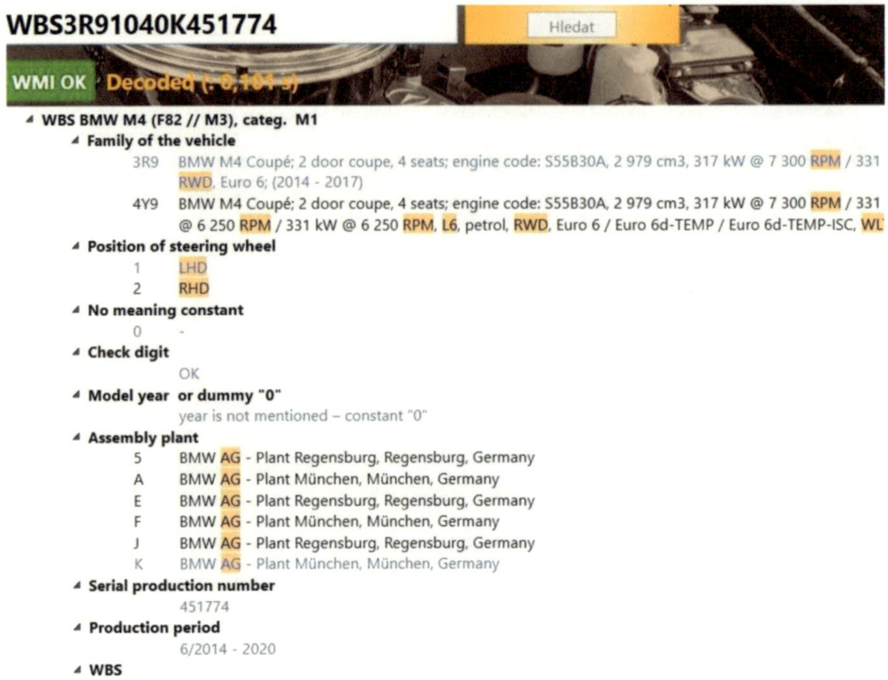

Figure 3.
Example of VIN decoding of a BMW vehicle using the VINexpert application. In the left column, there is a list of allowable values for every VIN position for specific logical sections of the identifier (family of the vehicle, position of steering wheel, etc.). Source: Roman Rak.

- **VDS—Vehicle Description Section.** The section specifies the technical and other characteristics of the vehicle and their structure, and coding depends on the vehicle manufacturer. The ninth position of the VIN (i.e. the sixth position of the VDS) can be used for so-called check digit mechanism. If the check digit mechanism is used, a globally defined standard is used. This mechanism is mandatory in the USA, but not in other countries. Again, it depends on the voluntariness and the will of the manufacturer whether or not to use the mechanism. The check digit then determines whether there was an error in the VIN entry (copy). The check digit mechanism is very important to ensure data quality in any information system.

- **VIS—Vehicle Identifier Section**. This section always contains the vehicle sequence and serial number. At the manufacturer's discretion, the year of manufacture, or so-called model year and factory, or other information (especially for vehicles produced in small batches) may be included there.

A data sample of 4,059,009 records from the Vehicle Inspection Register managed by the Ministry of Transport of the Czech Republic was used for the VIN quality analysis. The data sample is valid as of 31 December 2018.

The VIN distressed vehicle in a fraction of a second, without having to link to the various national registers of every EU member state.

The VINdecoder application is based on special algorithms and knowledge database. This database contains VIN structures and related vehicle information.

The database holds all the VIN information on a global scale, defined by vehicle manufacturers since 1986, when the VIN was defined as a worldwide unique identifier and made mandatory in automotive practice. The knowledge database contains approximately 8,900 basic type models for decoding the VIN structure of vehicles operating in the European Union. The knowledge database thus analytically covers 99.8 % of registered vehicles (including motorcycles, trucks, buses, tractors, semi-trailers, trailers, work machines, etc.).

Quality of this sample was checked using the special VINdecoder application (product name VINexpert). This application was originally designed and is still used in the Czech and Slovak Republics to decode the basic information contained in the VIN of a specific vehicle for the needs of the integrated rescue system within the framework of the pan-European eCALL project in cases where a vehicle has crashed and sent a distress signal [18], containing, among other things, its VIN identifier [19, 20].

The VIN of the crashed vehicle is transmitted from the vehicle using the network of mobile telephone operators and subsequently decoded. This provides the emergency services with basic information about.

2.1 Factors affecting the quality of VINs in information systems

There are two main factors, direct (objective) and indirect (subjective), that determine how well a VIN is entered into a computer application:

- Technology of entering the VIN in the information system
- Control mechanisms for verifying the VIN reality and correctness

2.2 Technology of entering the VIN in the information system

This is a way to technically enter the VIN into a computer application and its database. Historically, the methods of entry and the possibilities of its execution have changed, depending on the evolution of technology, computer technology, and peripherals [21]. There are the following three basic options for entering VINs into records dealing with motor vehicles:

- Manual entry (RT1[1], see **Figure 4**)
- Opto-electronic entry (RT2)
- Digital entry from vehicle control units (RT3)

The entry technology is considered to be a direct, objective factor affecting the quality of VINs in information systems. The VIN is always entered into computer applications using a certain technology, which can include a simple manual copy from the document submitted with the vehicle [22].

[1] RT – Reading Technology

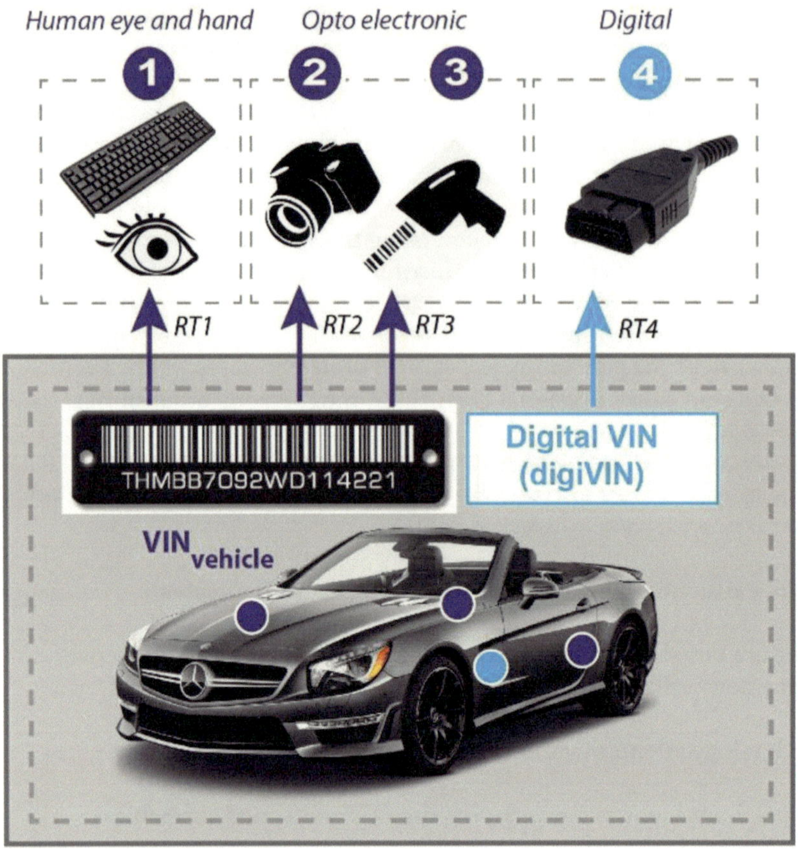

Figure 4.
Various technological principles for reading the VIN from a vehicle or documents and then entering it into an information system that processes vehicle data. Source: Roman Rak.

2.2.1 Manual entry

The VIN is entered by the user using a keyboard from its paper template, from the vehicle registration document, COC[2] sheet, etc. Manual entry is still the most common way of capturing data in vehicle registration information systems. This method features the lowest quality of data captured, i.e. VIN error rate. Incorrect, unrealistic VIN values may be shown in the documents. During the copying process, the user may make unknown, unintentional errors (misreading the VIN entry from the document, typing it incorrectly, swapping adjacent VIN characters due to finger motor skills on the keyboard, etc.). Deliberate errors in the VIN entry in order to change the identity of the vehicle, so that it cannot be checked against, for example, police tracking systems containing stolen vehicles, are quite common as well.

2.3 Opto-electronic entry

An opto-electronic interface (peripheral) is connected to a computer application working with vehicle registration, which ultimately makes an electronic entry of the

[2] COC – Certificate of Conformity

VIN without the need to use a keyboard. The interface can be a digital camera with special software that converts the visual (image) form of the VIN into an electronic, text form; a 2D or 3D code scanner or an OCR reader to extract the text content from the vehicle document. The camera or scanners can be used to capture the VIN from the homologation or data plates of the vehicle, from the VIN located under the windscreen or from the VIN physically stamped into the vehicle body.

This procedure of VIN acquisition guarantees a high-quality VIN. Unintentional errors in the VIN due to human fatigue, inattention, and inability to read or write correctly are excluded in this case. However, deliberate errors cannot be completely excluded, where a person deliberately makes a VIN using opto-electronic peripherals from another document, vehicle, etc. In order to exclude this type of error, a series of additional photographs are taken of the overall object from which the details were taken.

2.3.1 Digital entry from the vehicle control units

The VIN is transferred from the vehicle control unit, using a standardized OBD II interface, directly to the vehicle registration application, completely automatically, excluding any human factor. The human role is only to connect the connector to the vehicle interface at a standardized location in the driver's workstation area [23]. The data transfer is usually implemented wirelessly [24, 25].

This procedure eliminates both intentional and unintentional human errors. There are currently on average over 80 electronic control units (ECUs) in a modern vehicle (*Electronic Control Unit*). Many of these contain a digital VIN or other identifier [26, 27]. In addition, differences in VIN values may reveal unauthorized fitting (replacement) of major vehicle components [28], which may even come from illegal activities—stolen or scrapped vehicles, etc. [29, 30].

2.4 Control mechanisms for verifying the VIN reality and correctness

The basic task when entering a VIN into a computer application is to ensure it features no errors, i.e. its necessary data quality. This is due to the fact that the VIN is the basic identifier of the vehicle and also the linking, primary key among various databases.

Verification mechanisms for checking the reality and correctness (error-free) may or may not be used in practice. It is up to the responsibility and knowledge of the owner of the vehicle registration system whether the checking mechanisms are implemented and whether they insist on their application in daily practice without any exceptions. We are talking about an indirect, subjective factor influencing the final quality of VINs in information systems.

The most important vehicle register, which also serves as a reference for other information systems and related processes, is the national vehicle register. This register exists in every country and is usually under the responsibility of the Ministry of Transport (or other similar institution), exceptionally under the Ministry of the Interior. As this is the national reference register from which information on the vehicle (its owners, operators, technical condition, etc.) is taken, the quality of the VIN must be absolutely perfect.

Before its actual entry, the VIN can be entered into the registry database, and a number of logical checks can be carried out automatically to confirm that the registered vehicle is in order – not stolen, not searched for, etc.

In the EU countries, when entering a VIN in the national vehicle register, information is checked in particular in:

- The national police records of stolen vehicles (see **Figure 5**), [1];
- International police Schengen vehicle registration [2];
- International police records of stolen vehicles of the EU member states [3];
- National vehicle registers of the EU member states using the EUCARIS interface [4].

Searching for a vehicle via its VIN in police records has one basic specificity that we must always keep in mind: the fact that we cannot find the vehicle we are looking for in police records (national and international) on the basis of its VIN does not mean that the vehicle is OK! We must take these facts into account:

1. The vehicle owner (who is on holiday by air, for example) has not yet discovered that his vehicle has been stolen and, therefore, could not report the loss to the police. The vehicle might not have entered the police's search systems.

2. The perpetrator, a well-organized gang, transported the stolen vehicle from one country to another and registered it abroad within a very short time (hours). Organized gangs are quicker than police processes, so the transfer of information

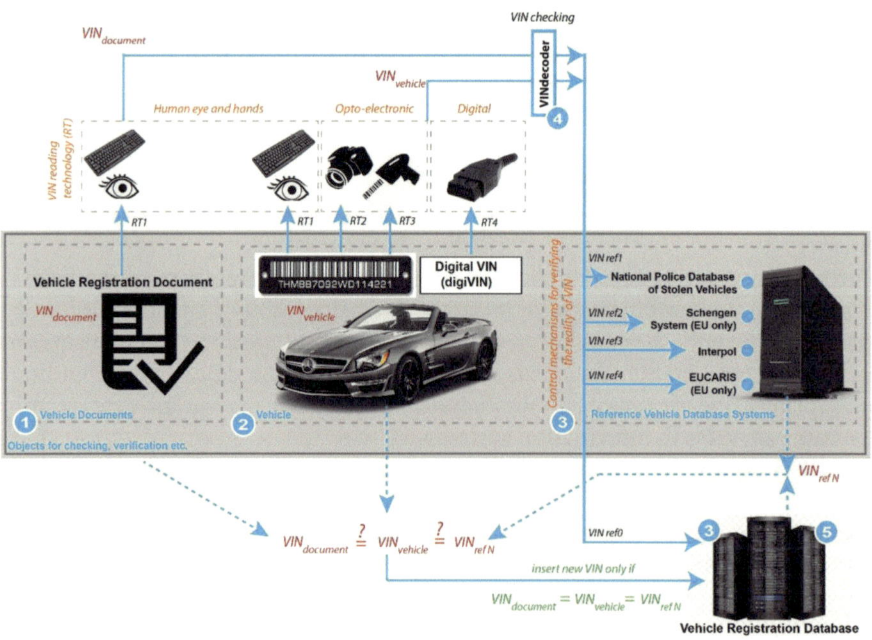

Figure 5.
Basic principle of reading the VIN from the vehicle and subsequent control mechanisms for verifying the reality and correctness (error-free) of the VIN in various information systems (VINref1 – VINref4). Source: Roman Rak.

about a stolen vehicle among any national and international search records will not take place in time.

3. The perpetrator deliberately changes the vehicle identity (its VIN), or the user sending the VIN for checking in police (and other IS) makes an error in the description, artificially, inadvertently creates a VIN of another vehicle, so that he changes the vehicle identity and in response receives information about a completely different vehicle.

4. The vehicle owner is part of an organized criminal group. The owner sells his vehicle abroad himself or through intermediaries, where he waits for the new owner to register it in a national vehicle register. The reference checking mechanisms in the police information systems do not work, because the vehicle is not yet reported as stolen. It is only after the vehicle has been successfully registered abroad that the perpetrator reports the vehicle as stolen in their home country and fraudulently obtains the insurance amount from the insurance company.

In all of the aforementioned cases, a query to police information systems results in the erroneous information that the vehicle is not searched for and has not been stolen.

2.5 The issue of the complexity of interconnected information systems

Today's era is characterized by a very dynamic exchange of data/information that is essential for correct and timely decision-making. Vehicle data is stored in various information systems so that it is necessary to link diverse information systems to obtain a comprehensive picture of the overall situation (**Figures 6** and 7). In the case of vehicles, the linking key is the VIN. The VIN is globally unique and is physically located on the vehicle at several standardized locations so that it is always possible to link the physical identity of the vehicle to its identity in the information systems. The quality of the VIN entry in each information system separately then determines the searchability of all the interlinked information systems.

2.6 What is the quality (Error Rate) of the VIN in the information systems?

This was a fundamental question that was one of the main objectives of the research conducted. The research was carried out on a data sample of 4 million vehicles from the Vehicle Inspection Register managed by the Ministry of Transport of the Czech Republic. This system is characterized by the fact that no checks are carried out when a vehicle roadworthiness test is entered. The VIN is copied manually from the vehicle document presented for technical inspection, without any check in the information system against any previous inspections of the same vehicle, without any technical means of obtaining the VIN from the vehicle or its documents. Nor is there any check of the existence of the vehicle being checked against the national vehicle register. There is also no check on the formal, logical structure of the VIN (length of the VIN string and prohibited characters) and no calculation of the check digit either.

Using the VINdecoder application (VINexpert), every record of the 4 million record sample set was batch analyzed. Each VIN of all the vehicles in the Czech Vehicle Inspection Register was examined and evaluated for its decodability and

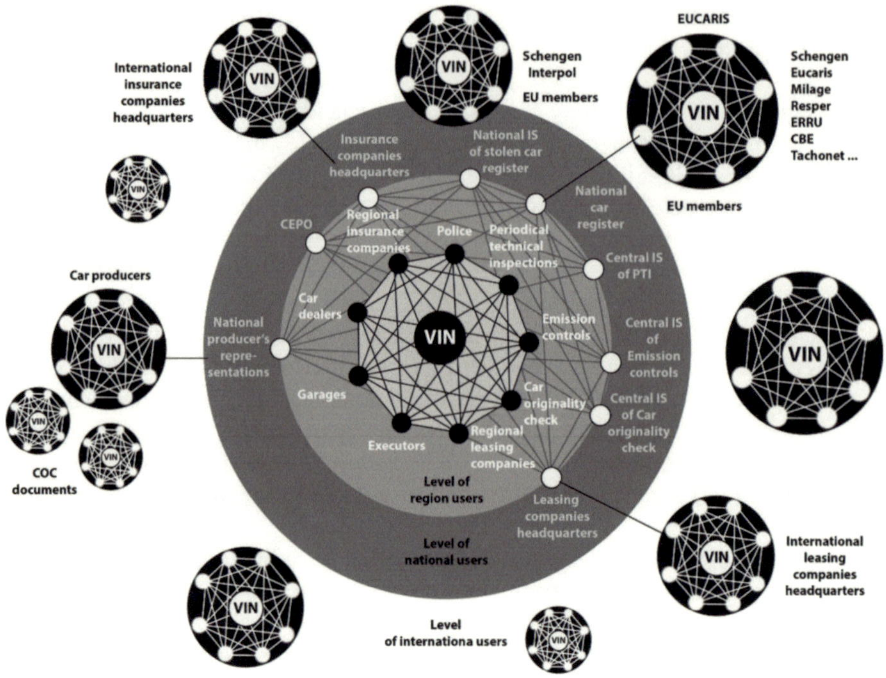

Figure 6.
The possibilities of linking vehicle data between different information systems, both nationally and internationally, can be seen as complex networks of different links, the basis of which is an error-free VIN. Any error (intentional or unintentional) means a failure in the efficiency and comprehensiveness of the vehicle information provided. Source: Roman Rak.

WBA3R91040K451754

WMI NOT OK Decoded with checksum error (: 0,044 s)

WBA BMW M4 (F82 // M3), categ. M1

Family of the vehicle	BMW M4 Coupé; 2 door coupe, 4 seats; engine code: S55B30A, 2 979 cm3, 317 kW @ 7 300 RPM / 331 kW @ 7 000 RPM, L6, petrol, RWD, Euro 6; (2014 - 2017)
Position of steering wheel	LHD
No meaning constant	-
Check digit	NOT OK
Model year or dummy "0"	year is not mentioned – constant "0"
Assembly plant	BMW AG - Plant München, München, Germany
Serial production number	451754
Production period	6/2014 - 2020
WBS	BMW M GmbH, Mnichov, Germany (passenger car) (correct)
WBA	BMW AG, München, Germany (passenger car, SUV) (incorrect)
WMI NOT OK	Check WMI if is written correctly! If yes then vehicle was identified on VDS section of another make. There is no data in application. Ignor this error.
Last global homologation	e1*2007/46*0377*18
Destination region	Europe

Figure 7.
Example of display of errors found in a vehicle identifier. Source: Roman Rak.

checked for correctness. If any decoding was incorrect, the causes were sought. The final report featured the statuses for every record listed in **Table 1**.

3. Discussion

We live in the age of modern information technology. Therefore, it should be fully assumed that we will work with accurate, correct, up-to-date data, which is stored

Status number (StN)	Status description	Status remarks
1	VIN decoded OK; VIN is used in the vehicle manufacturer's structure after 1986	
2	VIN decoded OK; VIN is used in the vehicle manufacturer's structure before 1986	Especially the German vehicle manufacturers (Audi, BMW, etc.)
3	Error in the VIN structure, VIN not decoded	The analyzed VIN does not correspond to the common standards and values used in practice.
4	VIN decoded, but the calculation of the check digit is incorrect	The user entering the VIN into the computer database made an error, confusing one or more characters with incorrect characters.
5	Unknown WMI	The user made an error in the first three characters of the VIN and created a nonexistent, unapproved combination for WMI
6	Correct VIN length (VIN string[1] contains exactly 17 characters). There are forbidden characters in the VIN.	Forbidden characters: O, Q, I. The user has confused the 0-O, J-I, and 0-Q characters, etc.
7	Incorrect VIN length. However, there are no forbidden characters in the VIN.	The length of the VIN string is less than or greater than 17 characters.
8	Incorrect VIN length. The VIN contains prohibited characters as well.	The length of the VIN string is less than or greater than 17 characters.

Source: Authors[1]Translator's note: string = string (computer terminology)

Table 1.
Summary of basic VIN statuses identified during the correctness analysis.

in information systems for various purposes—production, administrative, control, public transparency and efficiency, health and property protection, security, etc. It should also be assumed that data processing, including its acquisition (primary entry) into information systems in the public administration, will be supported by automated processes and modern technological means, which are already in common use in our everyday practice.

However, this is not the case in practice in the performance of state agencies (registers) working mainly with motor vehicles. There are objective and subjective reasons for this. The issue of vehicle registration as such is very specific in that there are a number of diverse manufacturers creating completely new products and global standardization in some areas (also related to registration practice) is not sufficiently flexible and at the same time not consistently observed. On the one hand, there is technological, major globalization in terms of technical or consumer aspects; on the other hand, the relevant legislation or standardization is also intended to be global (or at least pan-European), but its implementation is delayed by many years due to the adoption and implementation of European directives and regulations in national legislation.

In practice, the basic purposes of any systematic creation of various records and registers are often forgotten. In the past, every object entered into information systems was usually physically checked to ensure the quality of the information, especially key identifiers and other important characteristics.

The current registration processes, specifically in the case of motor vehicles, are only a "paper" matter as they are formally separated from each other. Vehicle registration is based only on the documents submitted, and there is a relatively large margin for error or even deliberate manipulation. In other words, one cannot, for example, technologically read the VIN by optically scanning its physical stamping from the vehicle body, read the VIN digitally from the vehicle control units, or scan the barcode. In the registration process, the vehicle is not physically present at the place of registration. This issue can theoretically be solved by carrying out quality technical inspections and vehicle originality checks that physically take place elsewhere and at different times. Unfortunately, even here the potential of opto-electronic or electronic (digital) technologies that are naturally available cannot be effectively used, because there is no standardized support for the uniform use of barcode or other technologies for recording VINs on vehicles, reading digital VINs from vehicles by manufacturers [31]. Not everyone uses barcodes or QR codes. There is no uniform device for reading (extracting) digital VINs from vehicle control units today that is capable of reading these values in general from all models that are at least simultaneously produced. For every manufacturer (manufacturing concern), it is necessary to have its proprietary technology available, which is not possible in independent inspection practice.

The basic research results obtained, presented in **Tables 1–3**, correspond to the practice of manual data acquisition and, at the same time, the lack of understanding of the seriousness of the vehicle identification issue in the design of the information system. The analysis shows that within the VIN item, 8.79 % percent of records do not

Status number (StN)	Status description	Decoded	Number of records	%
1	VIN decoded OK; VIN is used in the vehicle manufacturer's structure after 1986	OK	3,701,777	91.19
2	VIN decoded OK; VIN is used in the vehicle manufacturer's structure before 1986	OK	731	0.02
3	Error in the VIN structure, VIN not decoded	Err	129,911	3.2
4	VIN decoded, but the calculation of the check digit is incorrect	Err	2,975	0.07
5	Unknown WMI	Err	1,611	0.04
6	Correct VIN length (the VIN chain has exactly 17 characters). There are forbidden characters in the VIN.	Err	1,501	0.04
7	Incorrect VIN length. However, there are no forbidden characters in the VIN.	Err	168,273	4.15
8	Incorrect VIN length. The VIN contains prohibited characters as well.	Err	52,230	1.29
	Summary:		**4,059,009**	**100**

Source: Authors

Table 2.
Overview of error types in the VIN identifier. Decoded – the VIN identifier is error-free (91.21 % overall); Err – 8.79 % of VINs in the database are erroneous. Err – decoding errors.

Value of the record in the VIN entry	Comment
0/012447V	Completely different values are entered in the VIN entry of the administrative nature and unrelated to the vehicle characteristics
353-016784	
GM51B-115745	The serial number of the individual vehicle unit, the user is not familiar with the concept of VIN and enters completely different serial numbers.
4150417	Only the last characters of the VIN, resembling the serial number used before 1986, are copied from the VIN.
03072844	
RC422202787	
TK9205TTTR2BP31083	One extra "T" character is entered in the VIN entry when typing.
TNK52012024	The user or clerk writes only the first part of the VIN, not the entire 17-character string.
UU2TAQB02768	

Source: Authors

Table 3.
Example of typical erroneous VIN entries. Users either write completely different information values into the database, unrelated to the VIN, or they copy only the first or last parts of the VIN.

correspond to ISO standards imposed on this identifier. In other words, the error rate for the key identifier VIN is almost 9 %; i.e., one in 11 vehicles is problematic in terms of its unambiguous identification.

A closer analysis reveals that 5.48 % (4.15 + 0.04 + 1.29; see **Table 3**) of all the VINs are incorrect. This is due to the incorrect identifier length (different from the 17 standard characters) and the use of prohibited characters O, Q, and I. These characters must not be used in the VIN structure in order to avoid optical confusion of character pairs such as 0-O, 0-Q, I-J, and I-1, because then the object of interest cannot be found correctly in the search. The analysis also shows (see **Table 3**) that, in practice, data is entered in the VIN entry, items which have a completely different predictive value and certainly do not belong in the VIN entry. Clerks enter various official numbers and file marks. In numerous cases, this includes shortening (front or back) the VIN, usually to only 6–8 positions, because the official thinks that this sequence (reminiscent of the pre-1986 body serial number entries) is sufficient to identify a vehicle. This issue is trivially solvable at the level of information system design because it is sufficient to check the length of the VIN identifier for 17 positions and for the forbidden characters O, Q, and J. Records that do not meet these criteria must be brought to the attention of the information system operator, and such records must not normally be entered into the computer database. This type of error is of an objective nature (incorrect design of the functionality of the information system) and can be corrected retrospectively at minimal cost so that further errors do not occur.

The analysis also shows that an additional 3.31 % of all the VINs are erroneous, and the errors are due to human factors [32–34], in particular fatigue, inattention of an unintentional nature and possible fraudulent behavior [35] to change the vehicle identity [36]. A single character of the 17-digit VIN can be mistyped or misspelled, and a new "artificial" or fictitious VIN is created, which either does not formally exist or belongs to a completely different vehicle. These errors can only be eliminated by using the check digit mechanism in the VIN and/or by checking the inserted VIN using so-called VINdecoders which check the VIN structure. As such, the VIN check digit mechanism only works in full if vehicle manufacturers in a given country are legally obliged to have this mechanism built into the vehicles they sell. An example is the USA. In Europe and other continents, there it is then necessary to use suitable VINdecoders that operate in real time.

4. Conclusions

The correctness and factual correctness of key object identifiers (e.g. motor vehicles) is one of the basic prerequisites for the functionality of any information system [37]. If this unique identifier is not correct (it is incorrect), then it is impossible to find an unambiguous result in any record (computer database) by a simple, single query [38].

Is the 9% error rate in a key (vehicle) register acceptable? From our professional point of view, it is not. We must also be aware that information systems are today interconnected, especially the state ones (e.g. the vehicle register administered by the Ministry of Transport and the register of stolen or interest vehicles maintained by the police). It can be assumed that the error rate in the VIN identifier exists objectively in all the information systems and is approximately the same. Thus, if, for example, two information systems are linked together by a VIN, the probability that the link will not occur is then twice as high, i.e. 18 %. The real world of linking records containing motor vehicles is very complex, as shown in **Figure 6**. But today, the key national registers of state and public administrations are no longer just a static matter for internal needs to "record something," but must also serve in an active mode for quick solutions, such as security threats of various natures. An example is the pan-European eCALL project to provide online information about a vehicle in distress (e.g. in case of a crash) from the vehicle register to the Integrated Rescue System forces for conducting rescue operations. The key identifier for the link is the VIN. If there is an error in it, it may mean that rescuers do not get the necessary information about the vehicle, its technical characteristics in time for their work, which may fundamentally affect the technology of intervention and, therefore, in certain cases endanger the health or lives of the accident participants [39, 40]. Similarly, counter-terrorism forces may not obtain information about the vehicle and its owner at critical moments [41, 42].

If we can eliminate trivial errors in the VIN caused by inaccurate entry of its length or the presence of the O, Q, and J forbidden characters in the VIN, the error rate still remains 3–5 %. This is due to human factors (inattention, fatigue, intent – fraud by which the vehicle identity is changed, etc.). Based on our research, the 3–5 % error rate generally applies to all the European countries where additional sophisticated checks on the formal and content accuracy of the VIN by decoding it using "VINdecoder" applications are not in place, and where the calculation using the check digit in the VIN cannot be simply applied. It has been noted that the error rate for private entities (banks, insurance companies, leasing companies, etc.) is significantly greater than in the government IS. In order to eliminate this type of error rate, three basic procedures can be recommended for the acquisition of data (VIN identifiers) into the information systems of the public administration: taking data directly from vehicle manufacturers in electronic form; multiple verification of vehicle identity between different information systems; and systematic use of the VINdecoder that checks the online VIN when it enters the information systems. The error rate of 5% is still high, because it doubles every time two information systems are linked, and this means in practice that one in 10 vehicles is practically, unequivocally unidentifiable! This is unacceptable for critical infrastructure information systems and must be addressed satisfactorily.

The VIN identifier is an important, unique key in all information systems [43]. The quality of the VIN, its flawlessness, decides whether the vehicle will be found during the search or not. Information systems are connected to each other precisely with the help of VIN. Among these information systems, in practice, there are also systems of the so-called critical infrastructure – search and registration systems of the police, security and rescue services, forensic institutions, etc. In other words, the VIN therefore decides whether a stolen vehicle will be found during an attempt to register it, whether fraud of a property nature will be prevented, or whether correct technical or personal data of the owner or operator of the vehicle will be provided in the course of rescue work when solving a traffic accident, etc.

In recent years, the VIN is also stored in the vehicle in its digital form. This allows the VIN to be better protected against its inadvertent or intentional changes. At the same time, the possibilities of eliminating VIN errors in various information systems are significantly improved. The VIN can be obtained using a standard OBD interface and then transferred directly to the relevant registration information system. This prevents errors caused by the human factor. At the same time, the VIN of each type of vehicle has its own internal, fixed logical structure that can be checked. In a similar way, the VIN is digitally stored in all electronic control units of the vehicle. Today's vehicle has an average of around 80–90 of these units.

It is therefore possible and desirable to automatically check the VIN in all these units, since the VIN must be identical everywhere. Different VIN values from one of the control units means unauthorized intervention in the control unit. This can be a warning signal, a suspicion that the vehicle comes from criminal activity or has been improperly handled outside of an authorized service center.

Acknowledgements

Our special thanks goes to Ambis University, especially its Department of Science and Research and the university's management, without whose financial support it would not be possible to publish the research results. Additional thanks goes to those providing general support to other research in the field of Security and Law.

Conflicts of interest

There is no conflict of interest.

Ethical approval

This article does not contain any studies with human participants performed by any of the authors.

Informed consent

This article does not contain any studies with human participants performed by any of the authors.

Data availability

Data sharing is not applicable to this article as no datasets were generated or analyzed during the current study.

Author details

Roman Rak* and Dagmar Kopencova
Department of Security and Law, AMBIS University, Prague, Czech Republic

*Address all correspondence to: roman.rak@ambis.cz

References

[1] Grigoryan VG. The new objects of forensic vehicle examinations. Theory and Practice of Forensic Science. 2019;**14**(2):84-91

[2] Wang Y, Ma S, Wei Y, Wang L. Vehicle-identification based on complementarity feature. In: 2020 Chinese Control and Decision Conference (CCDC). Hefei, China. 2018, 2020. pp. 3207-3211

[3] Chesnokova EV. Analysis of vehicle identification markings: Validation of methodologies for forensic purposes. Theory and Practice of Forensic Science. 2018;**13**(3):25-30

[4] Kopencova D, Rak R. Issues of vehicle digital forensics. In: 2020 XII International Science-Technical Conference AUTOMOTIVE SAFETY. Kielce; 2020. pp. 1-6

[5] Cruz-Jesus F, Oliveira T, Bacao F. The global digital divide: Evidence and drivers. Journal of Global Information Management (JGIM). 2018;**26**(2):1-26

[6] Kumar K, Assaf MH, Groza VZ, Petriu EM. Intelligent Parking Vehicle Identification and Classification System. In: 2021 IEEE International Workshop on Metrology for Automotive (MetroAutomotive). Modena, Italy: Virtual conference; 2021. pp. 157-162

[7] Chesnokova EV. A source of new information for vehicle identification. Theory and Practice of Forensic Science. 2017;**12**(3):85-88

[8] Zhavoronkov VA. Some peculiarities of information support for forensic analysis of vehicle identification markings. Theory and Practice of Forensic Science. 2018;**13**(3):31-37

[9] Rak R, Kopencova D. Actual issues of modern digital vehicle forensics. In: Internet of Thinks and Cloud Computing. China: Science Publishing Group; 2020. pp. 12-16

[10] Clady X, Negri P, Milgram M, Poulenard R. Multi-class vehicle type recognition system. In: Prevost L, Marinai S, Schwenker F, editors. Artificial Neural Networks in Pattern Recognition. Berlin, Heidelberg: Springer; 2008

[11] Rak R, Zrubak R. Project eCALL – Car in Emergency Situation. In: Horak R, Juricek L, Schwarz R, editors. 7th Scientific International Conference Crisis Management: Environmental Protection of Population – Conference Proceedings. Brno, Czech Republic. 2012. pp. 251-258

[12] Elagin V, Spirkina A, Buinevich M, Vladyko A. Technological aspects of blockchain application for vehicle-to-network. Information. 2020;**11**(10):465

[13] Mendiboure L, Chalouf MA, Krief F. Towards a blockchain-based SD-IoV for applications authentication and trust management. In: Skulimowski AMJ, Sheng Z, Khemiri-Kallel S, Cérin C, Hsu C-H, editors. IOV 2018. LNCS, vol. 11253. Cham: Springer; 2018. pp. 265-277

[14] Böhm K, Kubjatko T, Paula D, Schweiger H-G. New developments on EDR (Event Data Recorder) for automated vehicles. Open Engineering. 2020;**10**(1):140-146

[15] Matuszak Z, Jaskiewicz M, Ludwinek K, et al. Special characteristics of reliability for serial mechatronic systems. In: Selected Problems of

Electrical Engineering and Electronics (WZEE). Kielce, Poland. 2015

[16] Li L, Liu J, Cheng L, Qiu S, Wang W, Zhang X, et al. Creditcoin: A privacy-preserving blockchain-based incentive announcement network for communications of smart vehicles. IEEE Transactions on Intelligent Transportation Systems. 2018;**19**(7):2204-2220

[17] Jurecki RS, Jaskiewicz M. Analysis of road accidents in Poland over the last ten years. Scientific Journal of the Maritime University of Szczecin-Zeszyty Naukowe Akademii Morskiej W Szczecinie. 2012;**32**(104):65-70

[18] Khoshavi N, Tristani G, Sargolzaei A. Blockchain applications to improve operation and security of transportation systems: A survey. Electronics. 2021;**10**(5):629

[19] Balasubramaniam A, Gul MJJ, Menon VG, Paul A. Blockchain for intelligent transport system. IETE Technical Review. 2021;**38**(4):438-449

[20] Makridakis S, Christodoulou K, Blockchain. Current challenges and future prospects/applications. Future Internet. 2019;**11**(12):258

[21] MacHardy Z, Khan A, Obana K, Iwashina S. V2x access technologies: regulation, research, and remaining challenges. IEEE Communications Surveys and Tutorials. 2018;**20**(3):1858-1877

[22] Soltani S, Hosseini Seno SA. A formal model for event reconstruction in digital forensic investigation. Digital Investigation. 2019;**30**:148-160

[23] Mutawa NA, Bryce J, Franqueira VNL, Marrington A, Read JC. Behavioural Digital Fjorensics Model: Embedding Behavioural Evidence Analysis into the Investigation of Digital Crimes. Digital Investigation. 2019;**28**:70-82

[24] Corea F. The convergence of AI and blockchain. In: Applied Artificial Intelligence: Where AI Can Be Used in Business. Cham: Springer; 2019. pp. 19-26

[25] Mansor H, Markantonakis K, Akram R, Mayes K, Gurulian I. Log your car: The non-invasive vehicle forensics. In: 2016 IEEE International Conference Trustcom/BigDataSE/ISPA, Trust, Security and Privacy in Computing and Communications. Tianjin, China. 2016

[26] Jaskiewicz M, Lisiecki J, Lisiecki S, et al. Facility for performance testing of power transmission units. Scientific Journals of the Maritime University of Szczecin-Zeszyty Naukowe Akademii Morskiej W Szczecinie. 2015;**42**(114):14-25

[27] Dirnbach I, Kubjatko T, Kolla E, Ondruš J. Methodology designed to evaluate accidents at intersection crossings with respect to forensic purposes and transport sustainability. Sustainability (Switzerland). 2020;**12**(5):1972

[28] Le-Khac N-A, Jacobs DL, Nijhoff J, Bertens K, Choo RKK. Smart vehicle forensics: Challenges and case study. Future Generation Computer Systems. 2018

[29] Jánský J, Tušer I. Security management—Quantitative determination of crime risk in a selected region. In: Tušer I, Hošková-Mayerová Š, editors. Trends and Future Directions in Security and Emergency Management. Cham: Springer; 2022;**257**:30-45

[30] Kubjatko T, Görtz M, Macurova L, Ballay M. Synergy of forensic and security engineering in relation to

the model of deformation energies on vehicles after traffic accidents. Conference Paper. Transport Means - Proceedings of the International ConferenceVolume 2018-October. In: 22nd International Scientific on Conference Transport Means 2018. Trakai, Lithuania; 2018. pp. 1342, 26-1348

[31] Matuszak Z, Jaskiewicz M, Wieckowski D, et al. Remarks to the reliability assessment and to human actions. Especially Car Driver. In: 18th International Scientific Conference on LOGI Location. Lithuania: Klaipeda University; 2017. p. 134

[32] Tušer I, Hošková-Mayerová Š. Traffic safety sustainability and population protection in road tunnels. Quality and Quantity. 2020;**54**:1040-1055

[33] Tušer I, Hoskova-Mayerova S. Emergency Management in Resolving an Emergency Situation. Journal of Risk and Financial Management. 2020;**13**(11):262

[34] Kurilovska L, Hajdukova T. Taking advantage of information and communication technologies for teaching and raisin awareness in the fight against criminality during the COVID-19 Pandemic. In: Chova LG, Martínez AL, Torres IC, editors. INTED2021 Proceedings, 15th International Technology, Education and Development Conference. Valencia, Spain: IATED Academy; 2021. pp. 2394-2402

[35] van de Weijer SGA, Leukfeldt R, Bernasco W. Determinants of reporting cybercrime: A comparison between identity theft, consumer fraud, and hacking. European Journal of Criminology. 2019;**16**(4):486-508

[36] Raghavan S. Digital forensic research: current state of the art. CSIT. 2013;**1**:91-114

[37] Kolitschova P, Kerbic J. Forensic and technical aspects of vehicle identification labels. In: 11th International Scientific and Technical Conference on Automative Safety. Casta Papernicka, Slovakia; 2018

[38] Moravcik L, Jaskiewicz M. Boosting Car Safety in the EU. 11th International Scientific and Technical Conference on Automotive Safety Location: Casta Papiernicka, Slovakia. 2018

[39] Mendiboure L. Blockchain and Cooperative Intelligent Transport Systems: Challenges and Opportunities. In: Jemili I, Mosbah M, editors. Distributed Computing for Emerging Smart Networks. DiCES-N 2022. Cham: Springer; 2022

[40] Kopencova D, Felcan M, Rak R. Equilibrium and Limit States in Technical and Social Disciplines as Part of Risk Analysis and Management. In: Abdel WM, editor. Proceedings of 1st International Conference on Structural Damage Modelling and Assessment. Singapore: Springer; 2020

[41] Mostafa A. Vanet blockchain: A general framework for detecting malicious vehicles. The Journal of Communication. 2019;**14**(5):356-362

[42] Arshad H, Jantan AB, Abiodun OI. Digital Forensics: Review of Issues in Scientific Validation of Digital Evidence. Journal of Information Processing Systems. 2018;**14**(2):346-376

[43] Reedy P. Interpol review of digital evidence 2016 – 2019. Forensic Science International: Synergy. 2020;**2**:489-520

Chapter 7

Blockchain-Based Educational Certificates: A Proposal

Cristina-Maria Vipie, Alexandra-Diana Afumatu and Mihai Caramihai

Abstract

The development of science and technology is very fast and even increases year by year, especially in the education system. However, in practice, a lot of learning systems continue to employ traditional methods and thus, learning becomes a less active process. Thanks to the technological improvements that have taken place over time, the concept of Blockchain could be integrated in the educational sector. Although mainly used for digital currencies, it seems that it could significantly impact healthcare, pharmaceutical industry/drug supply chain management, and now, the educational sector. Hence, Blockchain is used by the majority of current systems as a secure mechanism for confirming and transferring student data, academic credentials, and databases of educational organizations. Thanks to Blockchain, which gives students a powerful tool to manage and share their learning success, employers can rely on trustworthy, realistic representations of students' potential based on academic achievement (trusted verification). This chapter aims to define the concept of Blockchain, both in general terms and within the educational system. It also aims to carry out a case study among students and establish correlations between their answers.

Keywords: Blockchain, educational certificates, students, Blockchain applications, educational sector, fake diploma problems, bitcoin, technology

1. Introduction

The advancement of science and technology is extremely rapid and even gets faster every year, particularly in the field of education. The most significant and cutting-edge technologies of the past 10 years are Blockchain and internet of things. A better understanding of Blockchain technology has made it possible for the education sector to use its comprehensive characteristics to the advantage of students, teachers, organizations, employers, and lifelong learners. The complicated field of document verification requires a number of difficult and time-consuming techniques to authenticate. The most significant records that universities offer to students are educational certificates. Fake certificates can be easily produced, though, because the issuance procedure is not very clear and verifiable [1].

Blockchain technology can be extremely important in this new environment since it can give systems the required foundation for secure and impenetrable operations. Blockchain is a digital ledger that stores and verifies data and information. In some institutions, Blockchain applications are currently being used for a range of purposes, including issuance and storage of certificates and diplomas, assessment of learning outcomes, support and management of academic degrees, protection of intellectual property, collaboration between students and their professors, learning accreditation, payment for studies, the creation of an academic passport (portfolio), and administration of the educational process [2].

2. Blockchain-based educational certificates, present, and future. Fake diploma problems

2.1 About Blockchain

2.1.1 What is Blockchain?

As it is written in Ref. [3], in 2008, Blockchain was introduced as a new technology. It began as a peer-to-peer database for recording Bitcoin cryptocurrency transactions [4]. The main objective, as it is said in Ref. [5], was to reduce any intermediaries and to allow clients to access their business directly, that is why Blockchain was created as a decentralized network of peer nodes. Each entity:

- can have a copy of the ledger of transactions;
- when it obtains agreement from the others in the network, it can create an item for its own repository;
- can verify that the ledger it possesses is identical to those throughout the network on a regular basis;
- broadcasts to the rest of the network any transaction made by its users.

The Blockchain is a distributed database or public ledger that stores a list of all digital events or transactions that have occurred and is fundamentally shared among all participants. Only if the majority of the participating parties decide, a transaction can be valid. Moreover, it is impossible to remove or change data unless with the agreement of all or the majority of the network participants, once it has been confirmed and inputted, exactly as it is said in Ref. [6],—an analogy in nontechnical terms; it is very easy for anyone to steal a cookie from a cookie jar that is kept in a hidden place than to steal from a jar that is placed in a market place, where thousands of people are keeping an eye on.

2.1.2 How does it work?

The name Blockchain is not by chance: A "chain" of separate "blocks" of data is often used to characterize the digital ledger. A new "block" is formed and attached to the "chain" as new data are uploaded to the network. This requires that all nodes update their Blockchain ledgers in order to be identical.

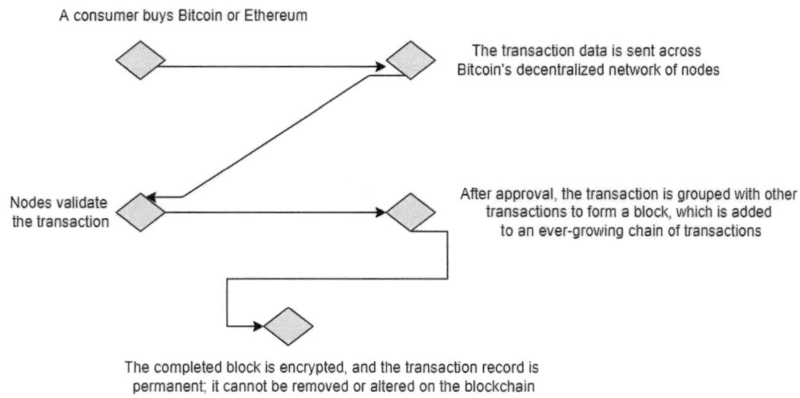

Figure 1.
An example of how Blockchain is used to verify and record bitcoin transactions [7].

The reason Blockchain is considered extremely secure is because of the way these new blocks with informations are formed. Before a new block can be added to the ledger, most nodes must verify and certify the legitimacy of the new data. An independent database or spreadsheet (the methods used before Blockchain), on the other hand, allows a person to make changes without supervision [7].

Figure 1 [7] shows an example of how Blockchain is used to verify and record Bitcoin transactions.

2.1.3 The importance of Blockchain

Blockchain has a number of qualities that make it useful. The success of Bitcoin, whose capital market presently stands at 191 billion USD [8], demonstrated the robustness and promising aspects of Blockchain. According to Ref. [6] and the UK government's office of research, Blockchain secures data records, lowers operational expenses, and increases transaction transparency.

The following are some of the interesting aspects, benefits, and significance of Blockchain, as it is mentioned in Ref. [6]:

- **Networked existence**: At the same moment, different users (nodes) on the Blockchain network store the same Blockchain data. Even if one node fails or loses data, other nodes in the network have a copy of the Blockchain and can continue to update it. The Blockchain can be recopied from other nodes. These feature guards against data loss, record tampering, and cryptocurrency costs unnecessary.

- **Decentralized nature**: Blockchain's decentralized nature eliminates the need for central authority and middlemen, making it more ideal for applications. Blockchain enables systems to be self-contained and devoid of the hazards that come with relying on middlemen and central authorities. Private Blockchains, on the other hand, can be partially or totally centralized while still benefiting from the other Blockchain features.

- **Data security and integrity**: Blockchain is secure in the sense that any changes to data in any block are discovered by a change in the block hash, which differs from the previously recorded hash in the next block. To be successful, a malicious user must change the block data for all computers on the network, which is essentially impossible in a large network. As a result, data on Blockchain is protected against alteration in this regard.

- **Traceability and transparency**: Because Blockchain records are time-stamped and saved on all complete nodes on the network, everyone can check and see all activity and transactions. All of a node's activity and transactions can be tracked if its address is known. Blockchain becomes visible and traceable as a result of this. It is also a useful platform for auditing and public services because it is suitable for fraud detection.

- **Efficiency**: Because middleman subsystems are removed, Blockchain allows systems to work autonomously with greater efficiency. This is one of the advantages that many companies and countries are hoping to obtain from Blockchain technology.

- **Verifiability**: The legitimacy of a record may be checked thanks to Blockchain's encryption. This may be difficult to do in other databases since it necessitates cryptographic technologies, such as Blockchain's digital signature.

- **Interoperability**: Blockchain provides a secure data-sharing platform that allows separate parties to share the same data and synchronize their services.

- **Cost savings**: Using Blockchain saves a lot of money because it eliminates the need for intermediary systems. Banks might save $20 billion each year if they used efficient Blockchain. One of the reasons why some banks and businesses seek to integrate Blockchain into their systems in order to save money is because of this economy.

Border control, government identity, insurance, shipping, real estate, advertising, waste management, energy, tourism, and a variety of other problems can all be solved with Blockchain technology. It is made up of numerous algorithms that are kept in the ledger and are used to detect faults. It also determines which block the error happened in.

2.1.4 Types of Blockchain

Due to the wide range of interests in Blockchain applications, the technology is divided into four categories: public, private, hybrid, and consortium Blockchains, as can be seen in the following **Figure 2**.

For a better view of these four categories of Blockchain, in the **Table 1**, there are the advantages, disadvantages, and some use cases of each one:

2.2 Applications of Blockchain

Blockchain was originally employed in cryptocurrencies, with Bitcoin being the first to demonstrate its success. There are numerous Blockchain applications today, as summarized in the following **Figure 3**:

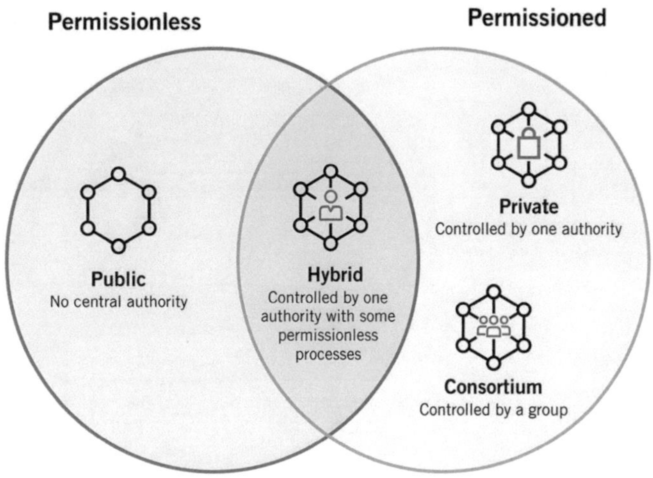

Figure 2.
Types of Blockchain [9].

	Public	Hybrid	Consortium	Private
Describe	No central authority.	Controlled by authority with some permissionless processes.	Controlled by a group.	Controlled by one authority.
Advantages	Transparency Independence Trust	Access control Performance Scalability	Access control Scalability Security	Access control Performance
Disadvantages	Performance Scalability Security	Transparency Upgrading	Transparency	Trust Auditability
Use case	Document validation Cryptocurrency	Medical records Real estate	Banking Research Supply chain	Supply chain Asset ownership

Table 1.
Differences between types of Blockchain [10].

To be more specific, the following list will explain a little bit about the importance of the use of Blockchain in the most important domains [6]:

- Cryptocurrencies;
- Smart contract;
- Healthcare management;
- Insurance;
- Banking and finance;
- IoT industry;
- Decentralized data storage;

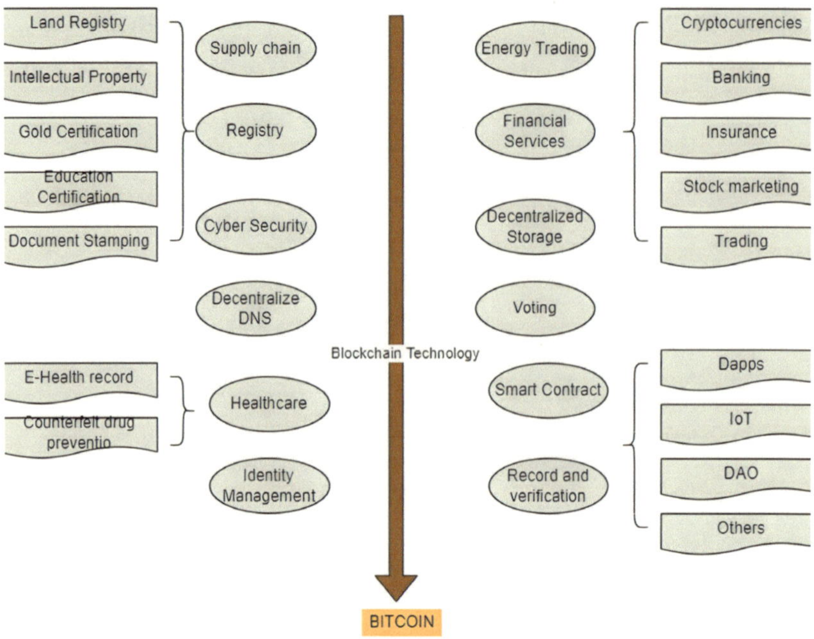

Figure 3.
Blockchain applications tree [6].

- Intellectual properties and document stamping;
- Digital identity management;
- Project management;
- Cybersecurity;
- Asset registry and tokenization.

2.3 Advantages and disadvantages of using Blockchain

According to Ref. [11], Blockchain technology has been portrayed as a disruptive technology that provides unprecedented levels of security, which is required and wanted not only by the IT and finance industries but by all industries in general, making it a very adaptable technology.

Despite its many advantages, Blockchain technology is far from flawless, and it has advantages and problems in its deployment, much like every revolutionary technology.

Advantages:

- Higher accuracy of transactions,
- No need for intermediaries,
- Extra security,

- Efficient transfers,
- Decentralization,
- Network distribution,
- Resistant and resilient.

Disadvantages:

- Limit on transactions per second,
- High energy costs,
- Potential for illegal activity,
- Risk of asset loss,
- Private keys,
- Unemployment,
- Storage.

And there are also some characteristics of the Blockchain network that can be both an advantage and a disadvantage, such as:

- Immutability of information,
- Anonymity.

2.4 Educational sector

The education sector is undergoing a technological transition; yet, there are numerous challenges with this. Virtual classrooms are taking over traditional classrooms. The real question is how might Blockchain technology help educational institutions and students' study more effectively [12].

This topic can be answered in a variety of ways; however, it will focus on three major segments to describe the advantages of Blockchain solutions and for whom it is beneficial:

- Educational institutions (universities, start-ups, and nongovernmental organizations) that are searching for solutions to improve the efficiency and security of student data storage and management;
- Learners who would benefit from more interesting, dependable, and long-term methods of accumulating, attesting, and sharing information;
- Employers who want to assess the validity of students' talents and credentials in a trustworthy and secure manner [12].

It is worth emphasizing that the application of Blockchain in academia is still in its early stages, which has an impact on the availability and quality of research on the subject. The majority of existing solutions employ Blockchain as a secure system for validating and distributing personal student data and academic diplomas, along with educational organizations' databases.

2.4.1 Blockchain technology for educational sector

After some research, it is clear to say the best way to explain how Blockchain could be implemented in the educational sector is to divide the interests, the same as it is explained in Ref. [12].

a. Enhancing security and efficiency for educational institutions, corporations, and students

    Blockchain technology has the potential to protect students' data by ensuring their identification, privacy, and security. As previously stated, Blockchain ensures integrity through its hash chain, which provides security and authenticity. Students, for example, cannot change past educational certifications stored on the Blockchain. Furthermore, because Blockchain does not store data, but rather a hash of it, privacy is assured. Before being stored on the Blockchain, the data might be encrypted if desired.

    As it is mentioned in Ref. [13], a variety of Blockchain-powered efficiency applications, including record-keeping applications, such as digital credentials and intellectual properties, simplifying diploma verification, and fast and reliable student payments. Not only do these technologies save money and time for educational institutions, but they also save time and money for companies and individual students.

b. Integrating trust and transparency

    Employers may be assured that job applicants have the required abilities to succeed in the industry, since Blockchain guarantees that students cannot falsify their grades, degrees, or certifications. Blockchain becomes a trust anchor of one truth for credentials|| as it is said in research "Tapscott and Kaplan, 2019."

    Moreover, this anchor gives job searchers and companies the option to make better matches. In general, because distributed ledger technologies enable learning and protect academic records, they improve interactions between "universities, companies, companies, and their relationships to society" by integrating trust and transparency into skill transactions and sharing procedures.

c. Learners' empowerment (self-sovereignty)

    The data (credentials, skills taught, etc.) linked with a student's identification is owned by the student, not by a central administration like a university. Students can keep their lifetime learning data (both inside and outside of the classroom), fully own it, and decide who has access to it.

Furthermore, even when students benefit from Blockchain "wallets," where they can store and share all of their learning data with various parties (students being complete owners of their identity-related data), they still benefit from the support of their professors, ensuring that they are not alone in their learning journeys.

Moreover, Blockchain can facilitate the most important element for educational institutes, such as

- Decentralization;
- Immutability;
- Smart contracts;
- Payment registry;
- Security;
- Transparency.

2.4.2 Blockchain-based educational certificates. Challenges reduced by Blockchain. Fake diploma problems

Students' educational qualifications become public and simply shareable with companies and universities for opportunities for future personal growth. Employers can base on reliable, realistic representations of students' potential based on academic accomplishment thanks to Blockchain, which provides students with an empowering tool to manage and share their learning achievement (trusted verification).

The following are some of the challenges that Blockchain can alleviate:

- It has the ability to establish an environment in which students' personal databases can be modified and subsequently stored. The Blockchain gives institutions access to data, which is a lot more precise, and any changes do not have to be as time-consuming;
- It can be seen in the development of an open-source environment. This can provide a place to store all of the documents that a student will need during his or her course of study, as well as provide an air of validity for students to carry less documents in their luggage, so offering an alternate mode of education;
- One key benefit of this technology to the educational backdrop is that it provides each student with a unique id, which aids students in matching up their information, and in the event of any project confusion between two students, it can be readily resolved. The ability to see the grades in real time can be a huge benefit;
- The issue that the schools/universities face is the high number of incidences of fraud and fraudulent diplomas offered to pupils. The fundamental objective of Blockchain technology is to get to a point, where every block is a proven block, so

that if there is any fraud, the necessary information is sent directly to the higher authorities, who may take swift action.

2.4.3 The advantages of using Blockchain in an educational environment

As it is mentioned in Ref. [6], the main advantages of using Blockchain in an educational environment are:

- It gives its customers the opportunity to check the veracity of a record against the Blockchain without involving the issuing organization.

- It works to eliminate the practice of issuing certifications on paper. Educational establishments' qualifications can be reliably and permanently recorded via Blockchain. Further developments could include automating certifications, transferring credits, or even preserving a complete record of a student's accomplishments throughout their academic career.

- Implementing Blockchain technology can drastically cut the costs incurred by educational institutions in maintaining data. It also enables the reduction of liability risks that are common while handling records.

- The records are open to the public and easily verified.

- Enhancing learners' interactivity.

- Supporting learners' career decisions.

- Improving the management of students' records.

- Enhancing trust.

- Identity authentication.

- Enhancing accountability and transparency.

- Better control of data access.

- Enhancing students' assessments.

- High security.

3. Analysis of the status quo of digital education recognition

3.1 Analysis of current state-of-the-art of the European digital education recognition using Blockchain in comparison to best practices in the world

As it is said in Ref. [14], the European Blockchain Partnership (EBP) was established after EU member states and Norway signed a declaration with the goal of providing digital public services that meet the requisite degree of digital security and maturity in today's society.

In future, all public services will use Blockchain technology. Blockchain is a great opportunity for Europe and member states to rethink their information systems, promote user trust and the protection of personal data, help create new business opportunities, and to establish new areas of leadership, benefiting citizens, public services, and companies. The partnership launched today enables member states to work together with the European Commission to turn the enormous potential of Blockchain technology into better services for citizens|| declared Mariya Gabriel, the commissioner for Digital Economy and Society, in 2018.

Many industries and colleges in Europe and beyond are becoming increasingly interested in Blockchain technology. Blockchain, a relatively new discovery in computer science, is a worldwide, cross-industry, and disruptive technology that is expected to drive global economic growth for another few decades.

When discussing a topic like Blockchain, it is natural to start with themes, such as technological transformation, digital economy, competency industries, and innovation system. This enables people to comprehend the context in which digital disruption occurs.

However, the socioeconomic forces that produce interest in technology (or alter in reaction to it) may be as essential, if not more so, than the digital technology itself. The most successful digital company concepts put people first and digital technology second. Within the educational setting, the phrase is quickly becoming associated with the ability of individual students to own, manage, and share information about their credentials without relying on the education sector as an authorized middleman.

Blockchain technology is perfect for securing, sharing, and verifying learning achievements as a new infrastructure. In the case of certificates, a Blockchain can maintain a list of the certificate's issuer and receiver, as well as the document signature (hash), in a public database (the Blockchain) that is replicated on thousands of computers all over the world [14].

3.1.1 Results of case studies on the application of Blockchain technology in education

As early Blockchain literature frequently refers to "self-sovereignty," or an individual's ability to own and control his or her own identity online, within an educational context, the term is quickly becoming synonymous with the autonomy of individual learners to own, manage, and share details of their credentials without relying on the education institution as a trusted intermediary [14].

3.1.1.1 Open University UK

Imagine a scenario where every learning activity is registered on the Blockchain, including informal learning, together with informal feedback. All assignment test scores will be mapped on learning environments across Europe. Europe-wide analytics could then be developed from the ground up. The best lecturers in Europe by the subject could be easily identified. Learning would become that much more interactive and reputations built on more tangible matrices||, declared Professor Domingue.

Professor Domingue suggests that the EU consider funding the development of an EU-wide Blockchain for educational experiments. Funding would be made available for more innovative projects on the same Blockchain. It should organize an education program as well as a series of informational meetings for various stakeholders. For example, colleges should use Blockchains to communicate with other colleges and universities and in different EU countries, fostering collaboration [14].

3.1.1.2 University of Nicosia

The University of Nicosia (UNIC) has declared a number of "world firsts" in its commitment to maximizing the potential of Blockchain in education. UNIC claims to be the first university to [14]:

- accept Bitcoin as payment for any degree program at the university since October 2013;
- has a course about cryptocurrency, called, Introduction to Digital Currencies|| since January 2014;
- offer an approved academic degree program in digital currency—a Master of Science in Digital Currency—taught online in English (from March 2014, with the first students graduating in June 2016);
- using its own in-house software platform, award academic certifications on the Bitcoin Blockchain since September 2014.

"It would be hugely valuable if high schools around the world had some common standard for accreditation and recognition. We cannot have 40 standards on a Blockchain. How does this become useful to higher education—which is being fed by secondary education? How can we get everyone to subscribe to the same standard? If any one institution like ours is doing it—it is limited; if a nation-state or all higher education institutions and schools in a country come on board—that would be very useful," claimed Mr. Polemitis.

In conclusion, it appears that Blockchain technology will likely be implemented by most EU member states. Some EU members will try to develop a national strategy for its use, and others are already testing specialized Blockchain applications.

Given the expense of implementing Blockchain technology, it is evident that, despite the excitement around the technology, it can only be applied to select use cases from a technical standpoint. As a result, a Blockchain-based application should only be used if it fits a specific set of requirements [14].

4. Methodology for collecting data

The further case study is based on finding out the perspective of students regarding the adoption of Blockchain technology in the education system. It is important to know the students' opinion, in order to find out their knowledge of Blockchain technology, but also the advantages and disadvantages that they consider in using such technology.

The research methods used are the following: a documentary study on the topic, secondary analysis of statistical data, and quantitative research by the method of distributing a questionnaire. The research method used is empirical, which is using primary quantitative research. This research was conducted on the basis of a structured questionnaire.

The questionnaire consisted of 10 questions, and its completion resulted in 147 responses from respondents. It was distributed online and was addressed to students at the Faculty of Automation and Computers, Polytechnic University of Bucharest. During the questionnaire, mainly quantitative questions were used (using the Likerd ordinal scale), but also qualitative questions were to find out the year in which the students first heard about the concept of Blockchain technology and in what context. Also, at the beginning of the questionnaire, questions were used to find out information about the respondents (gender and study program).

Quantitative questions (using the ordinal Likerd scale) were mainly used during the questionnaire. Respondents' responses were rated on a scale of 1 to 5 as follows: 1 = not important, 2 = low, 3 = medium, 4 = high, and 5 = very high.

The software used to analyze statistical data and find out the final results was SPSS (statistical package for the social sciences).

5. Analysis of data

At the basis of the work, there were five hypotheses that were analyzed. To summarize the results, five Figures will be presented (for each hypothesis). Each subpoint on the left will be represented by a color. On the right side, each subpoint will include the colors of the elements with which there is a correlation.

H1 - *The use of Blockchain technologies in the education sector influences the need for Blockchain knowledge in different fields.*

In order to be able to analyze H1, the 11 subpoints from question 5 (the level of use of Blockchain technologies in different cases in the educational system) and the 7 subpoints from question 6 (the level of knowledge about Blockchain technologies of different professions) will be taken into account. **Figure 4** shows the synthesis of hypothesis 1:

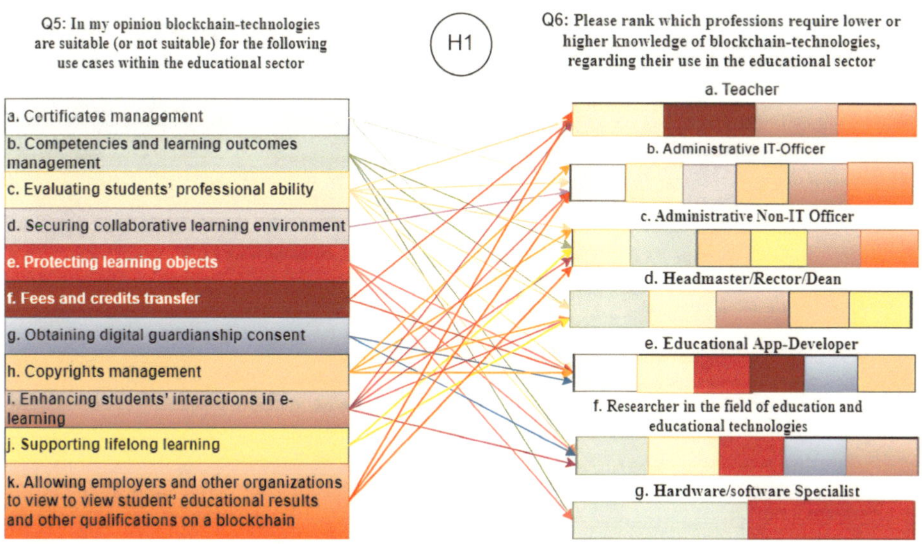

Figure 4.
Synthesis of hypothesis 1.

Between H1 elements, there were both weak positive correlations (statistically significant at a 95% confidence level) and reasonably positive correlations (statistically significant at a 99% confidence level). Thus, the following examples of *weak positive correlations* between the elements are mentioned:

- between the use of Blockchain technologies for certificate management and the need for high knowledge of the IT administrative officer,
- between the use of Blockchain technologies for securing a collaborative learning environment and the need for high knowledge of the administrative IT officer.

The following examples of *reasonable positive correlations* between the elements are mentioned:

- between the use of Blockchain technologies for protecting learning objects and the need for high knowledge of the researcher in the field of education and educational technologies,
- between the use of Blockchain technologies for fees and credits transfer and the need for high knowledge of the educational app developer.

H2 - *Issues to consider before including Blockchain technologies in the education sector influence the use of Blockchain technologies.*

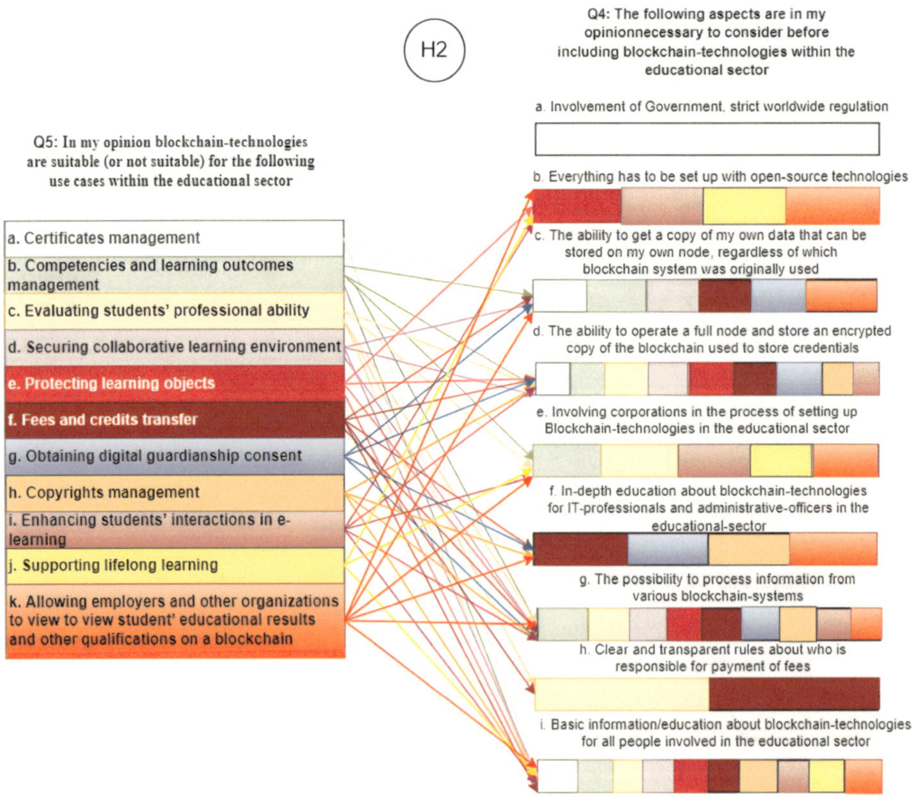

Figure 5.
Synthesis of hypothesis 2.

In order to be able to analyze H2, the 11 subpoints from question 5 (The level of use of Blockchain technologies in different cases in the educational system) and the 9 subpoints from question 4 (Issues to consider before including Blockchain technologies in the education sector) will be taken into account. **Figure 5** shows the synthesis of hypothesis 2:

Between H2 elements, there were both weak positive correlations (statistically significant at a 95% confidence level) and reasonably positive correlations (statistically significant at a 99% confidence level). Thus, the following examples of *weak positive correlations* between the elements are mentioned:

- between the use of Blockchain technologies for competencies and learning outcomes management and the need to consider "the ability to get a copy of data that can be stored on nodes, regardless of which Blockchain system was originally used" before including Blockchain technologies in the education sector,
- between the use of Blockchain technologies for securing a collaborative learning environment and the need to consider "the possibility to process information from various Blockchain systems" before including Blockchain technologies in the education sector.

The following examples of *reasonable positive correlations* between the elements are mentioned:

- between the use of Blockchain technologies for evaluating students' professional ability and the need to consider "the possibility to process information from various Blockchain systems" before including Blockchain technologies in the education sector,
- between the use of Blockchain technologies for protecting learning objects and the need to consider "basic information/education about Blockchain technologies for all people involved in the educational sector" before including Blockchain technologies in the education sector.

H3 - *Knowledge of Blockchain technologies in different professions influences the benefits of adopting Blockchain technologies.*

In order to be able to analyze H3, the 7 subpoints from question 6 (the level of knowledge about Blockchain technologies of different professions) and the 9 subpoints from question 7 (the benefits of adopting Blockchain technologies in education) will be taken into account. **Figure 6** shows the synthesis of hypothesis 3:

Between H3 elements, there were both weak positive correlations (statistically significant at a 95% confidence level) and reasonably positive correlations (statistically significant at a 99% confidence level). Thus, the following examples of *weak positive correlations* between the elements are mentioned:

- between the need for high knowledge of the teacher and better control of data access,
- between the need for high knowledge of the headmaster/rector/decan and enhancing learners' activity,

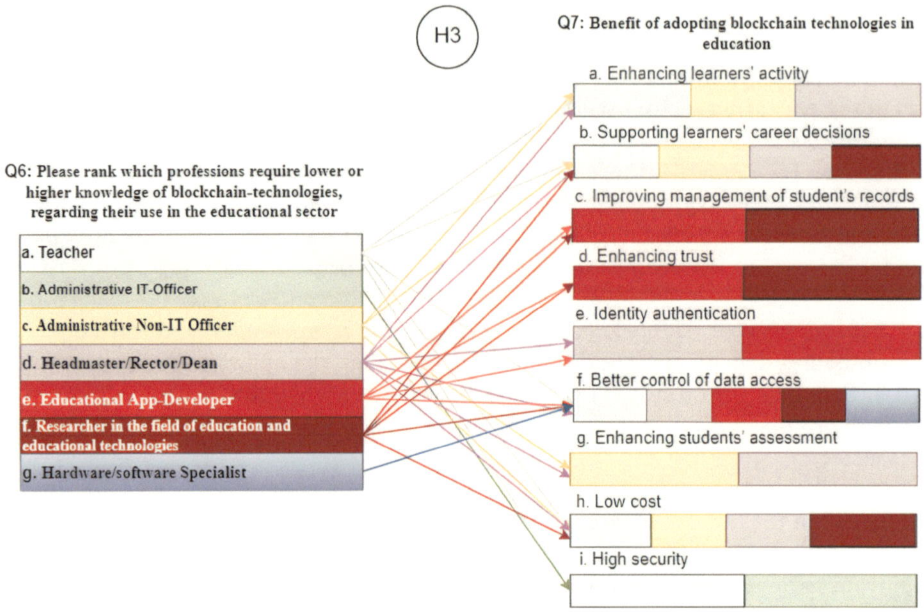

Figure 6.
Synthesis of hypothesis 3.

- between the need for high knowledge of the headmaster/rector/decan and identity authentication,
- between the need for high knowledge of the educational app developer and improving the management of students' records.

The following examples of *reasonable positive correlations* between the elements are mentioned:

- between the need for high knowledge of the teacher and the enhancing learners' activity,
- between the need for high knowledge of the teacher and the high security,
- between the need for high knowledge of the educational app developer and the enhancing trust,
- between the need for high knowledge of the researcher in the field of education and educational technologies and supporting learners' career decisions.

H4 - *Issues to consider before including Blockchain technologies in the education sector influence the benefits of adopting Blockchain technologies.*

In order to be able to analyze H4, the 9 subpoints from question 4 (issues to consider before including Blockchain technologies in the education sector) and the 9 subpoints from question 7 (the benefits of adopting Blockchain technologies in education) will be taken into account. **Figure** 7 shows the synthesis of hypothesis 4.

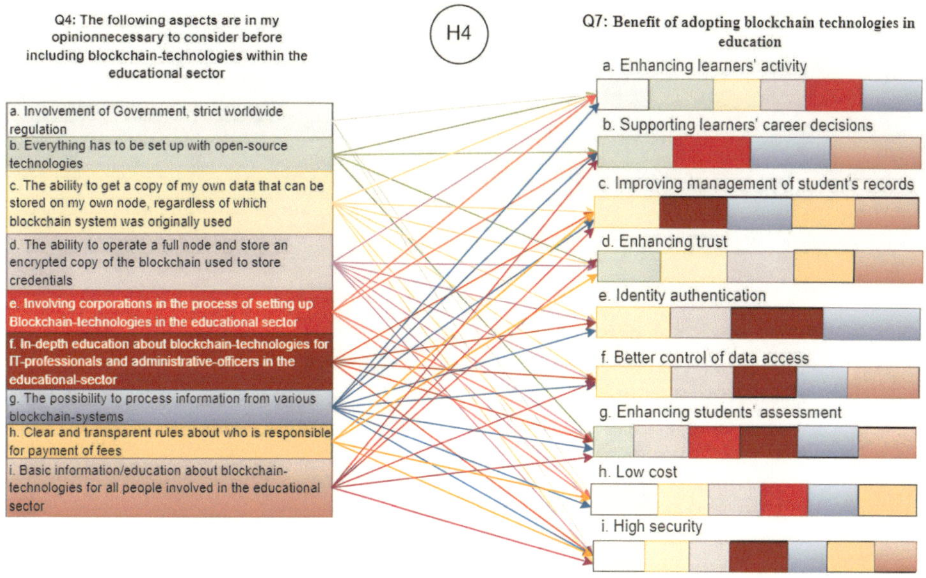

Figure 7.
Synthesis of hypothesis 4.

Between H4 elements, there were both weak positive correlations (statistically significant at a 95% confidence level) and reasonably positive correlations (statistically significant at a 99% confidence level). Thus, the following examples of *weak positive correlations* between the elements are mentioned:

- between the need to consider "involvement of Government, strict worldwide regulation" before including Blockchain technologies in the education sector and enhancing learners' activity,
- between the need to consider "everything has to be set up with opensource technologies" before including Blockchain technologies in the education sector and enhancing trust.

The following examples of *reasonable positive correlations* between the elements are mentioned:

- between the need to consider "involvement of Government, strict worldwide regulation" before including Blockchain technologies in the education sector and the low cost,
- between the need to consider "the ability to operate a full node and store an encrypted copy of the Blockchain used to store credentials" before including Blockchain technologies in the education sector and the enhancing trust,
- between the need to consider "the ability to operate a full node and store an encrypted copy of the Blockchain used to store credentials" before including Blockchain technologies in the education sector and the high security.

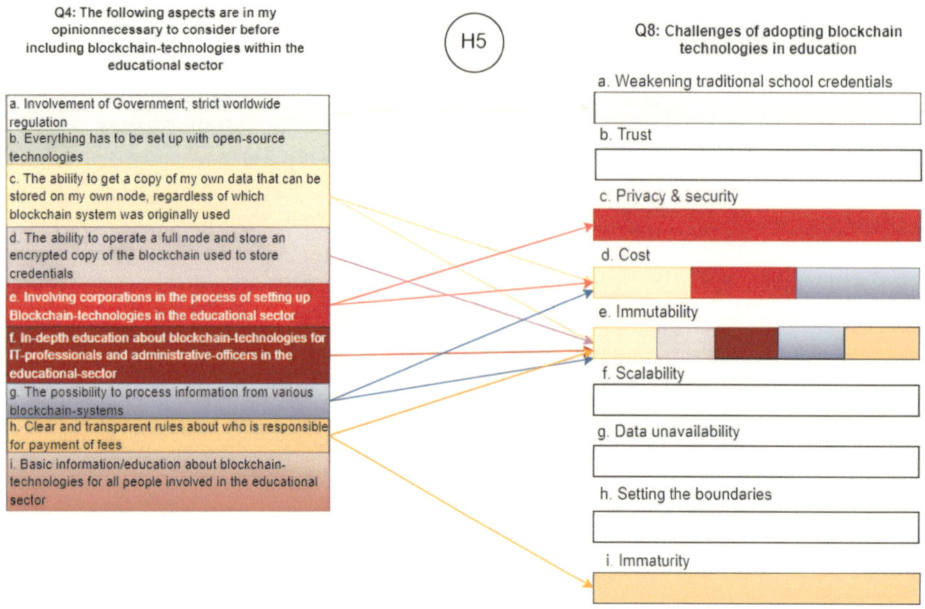

Figure 8.
Synthesis of hypothesis 5.

H5 - *Issues to consider before including Blockchain technologies in the education sector influence the challenges of adopting Blockchain technologies.*

In order to be able to analyze H5, the 9 subpoints from question 4 (issues to consider before including Blockchain technologies in the education sector) and the 9 subpoints from question 8 (challenges of adopting Blockchain technologies in education) will be taken into account. **Figure 8** shows the synthesis of hypothesis 5.

Between H5 elements, there were both weak positive correlations (statistically significant at a 95% confidence level) and reasonably positive correlations (statistically significant at a 99% confidence level). Thus, the following examples of *weak positive correlations* between the elements are mentioned:

- between the need to consider "the ability to get a copy of data that can be stored on node, regardless of which blockchain system was originally used" before including blockchain technologies in the education sector and immutability,
- between the need to consider "involvement of Government, strict worldwide regulation" before including Blockchain technologies in the education sector and weakening traditional school credentials,
- between the need to consider "involving corporations in the process of setting up Blockchain technologies in the educational sector" before including Blockchain technologies in the education sector and privacy and security,
- between the need to consider "the possibility to process information from various Blockchain systems" before including Blockchain technologies in the education sector and cost,

- between the need to consider "involving corporations in the process of setting up Blockchain-technologies in the educational sector" before including Blockchain technologies in the education sector and cost,
- between the need to consider "clear and transparent rules about who is responsible for payment of fees" before including Blockchain technologies in the education sector and immaturity.

The following examples of *reasonable positive correlations* between the elements are mentioned:

- between the need to consider "the ability to operate a full node and store an encrypted copy of the Blockchain used to store credentials" before including Blockchain technologies in the education sector and immutability,
- between the need to consider "the possibility to process information from various Blockchain systems" before including Blockchain technologies in the education sector and immutability,
- between the need to consider "clear and transparent rules about who is responsible for payment of fees" before including Blockchain technologies in the education sector and immutability.

6. Development of a pilot model using Blockchain concept for "record keeping" of students' degrees, certificates, and diplomas based on the previous analysis. Simulation of a case study

A pilot model using Blockchain concept for "record keeping" of students' degrees, certificates, and diplomas must take into account:

1. **Inputs:**
    - personal data (e.g., name, gender, date of birth, educational institution, etc.),
    - diplomas (e.g., engineering degree, etc.) and certificates (e.g., BA certification, etc.).
2. **Outputs:**
    - the hash of identity stored on Blockchain,
    - the hash of certificate stored on Blockchain.

Figure 9 shows a pilot model for "record keeping" of students' degrees, certificates, and diplomas.

Figure 10 shows the simulation of a case study.

Such a pilot model can bring numerous benefits for both students and institutions. The following benefits of the model are mentioned:

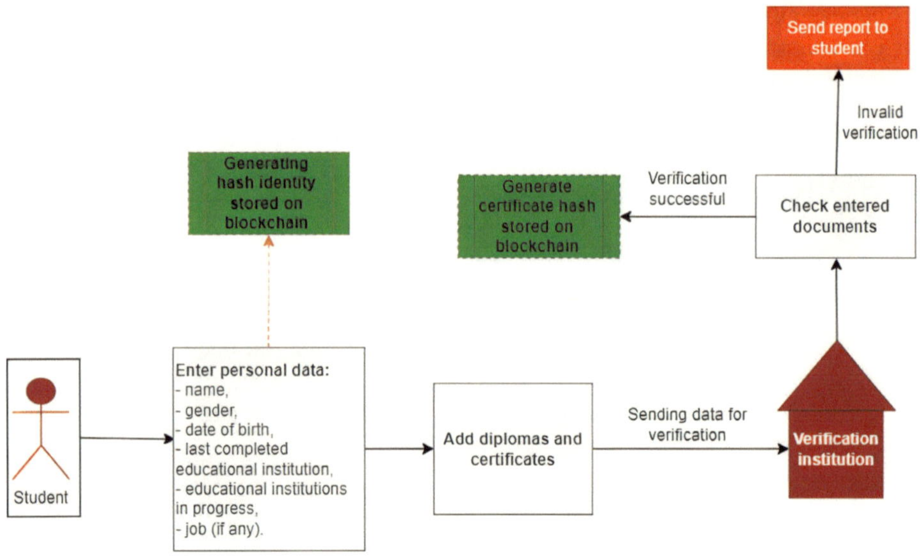

Figure 9.
Pilot model.

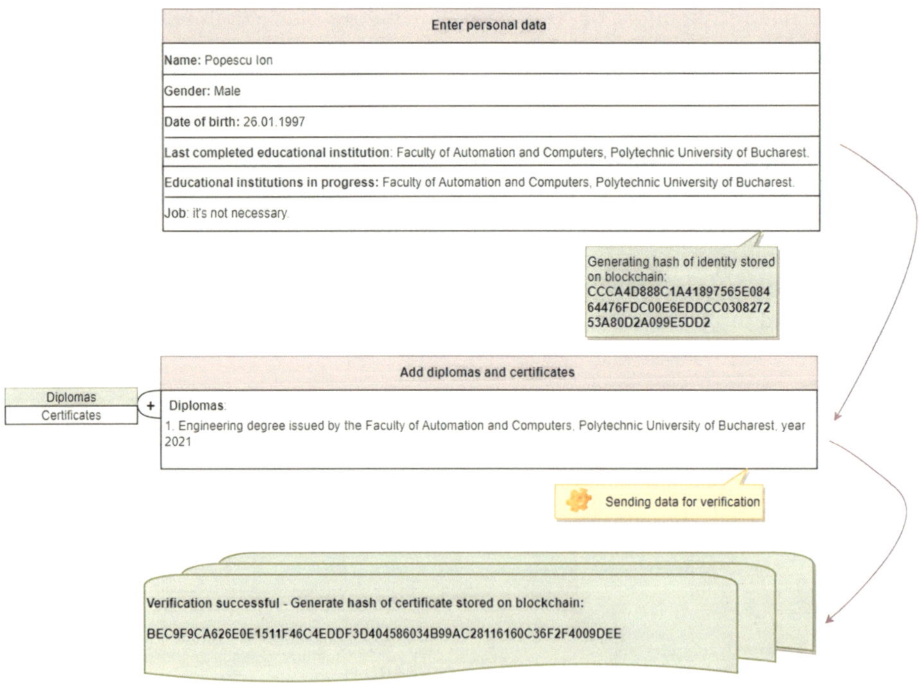

Figure 10.
Simulation of a case study.

- increased efficiency due to the simplicity of verifying the correctness of data and diplomas/certificates (reduced verification time),
- information confidentiality (due to hash generation),

- increasing confidence (the entered data cannot be falsified),
- protecting student data (by ensuring security with Blockchain technology).

In addition to the previously mentioned, students can use this data stored on Blockchain at the time of employment, increasing in this sense the trust in front of the employer.

7. Conclusions

The Blockchain is a distributed database or public ledger that stores a list of all digital events or transactions that have occurred and is fundamentally shared among all participants. Blockchain technology is perfect for securing, sharing, and verifying learning achievements as a new infrastructure. In the case of certificates, a Blockchain can maintain a list of the certificate's issuer and receiver, as well as the document signature (hash), in a public database (the Blockchain) that is replicated on thousands of computers all over the world.

Regarding the case study presented, it can be seen that all five hypotheses are validated. Thus, it can be said that the first link is between the use of Blockchain technologies in the education sector and the need for Blockchain knowledge in different professions. A second link is between the issues to be considered before including Blockchain technologies in the education sector and the use of Blockchain technologies. The third link is the knowledge of Blockchain technologies in different professions and the benefits of adopting Blockchain technologies. The fourth link is between the issues to be considered before including Blockchain technologies in the education sector and the benefits of adopting Blockchain technologies, but also a link between these issues and the challenges of adopting Blockchain technologies in the fifth hypothesis. Thus, the different subpoints that constitute the elements of the hypothesis are influenced by the existence of different reasonable correlations, but also weaker correlations.

Author details

Cristina-Maria Vipie, Alexandra-Diana Afumatu and Mihai Caramihai*
University Politehnica of Bucharest, Romania

*Address all correspondence to: m.caramihai@yahoo.com

References

[1] Siddiqui ST, Ahmad MO, Khamruddin M, Gupta AK and Singha AK. Blockchain and IoT for educational certificates generation and verification. 2022 2nd International Conference on Computing and Information Technology (ICCIT). 2022. pp. 298-303. DOI: 10.1109/ICCIT52419.2022.9711656

[2] Panagiotidis P. Blockchain in education - the case of language learning. European Journal of Education. 2022;**5**(1):66-83. DOI: 10.26417/443gjm83

[3] Alammary A, et al. Blockchain-Based Applications in Education: A Systematic Review. Applied Sciences. 2019;**9**(12): 2400. DOI: 10.3390/app9122400. Available from: https://www.mdpi.com/2076-3417/9/12/2400

[4] Nakamoto S. Bitcoin: A Peer-to-Peer Electronic Cash System. 2008. Available online: https://bitcoin.org/bitcoin.pdf

[5] Grech A, Camilleri AF. Blockchain in Education. Luxembourg: Publications Office of the European Union; 2017

[6] Sanka AI, Irfan M, Huang I, Cheung RC. A survey of breakthrough in blockchain technology: Adoptions, applications, challenges and future research. Computer Communications. 2021;**169**:179-202

[7] Rodeck D. What Is Blockchain?. Forbes Advisor. 28 Apr 2022. Available from: www.forbes.com/advisor/investing/cryptocurrency/what-is-blockchain/

[8] Cryptoreport. Live Crypto Prices and Trading. 2020. https://cryptoreport.com/all

[9] Wegrzyn KE, Wang E. Types of Blockchain: Public, Private, or Something in between. Foley & Lardner LLP. www.foley.com; 2021. http://www.foley.com/en/insights/publications/2021/08/types-of-blockchain-public-private-between

[10] What are the 4 Different Types of Blockchain Technology? SearchCIO, www.techtarget.com/searchcio/feature/What-are-the-4-different-types-of-blockchain-technology

[11] Advantages and Disadvantages of Blockchain. BBVA.CH. www.bbva.ch/en/news/advantages-and-disadvantages-of-blockchain/

[12] Blockchain in Education: Opportunities, Applications, and Challenges. Available from: https://firstmonday.org/ojs/index.php/fm/article/download/10654/9726

[13] Thayer TL. 4 Promising and Ambitious Blockchain Initiatives for Higher Education,|| Gartner Research, 2018. Available from: https://www.gartner.com/en/documents/3893582/4-promising-and-ambitious-blockchain-initiatives-for-hig

[14] Ali MA, Bhaya WS. Higher Education's certificates model based on Blockchain technology. Journal of Physics: Conference Series. 2021; **1879**(2):022091. DOI: 10.1088/1742-6596/1879/2/022091. https://iopscience.iop.org/article/10.1088/1742-6596/1879/2/022091

Chapter 8

Artificial Intelligence and the Media: Revisiting Digital Dichotomy Theory

Aondover Eric Msughter, Aondover Ogechi Perpetua and Adebija Lillian Itiafa

Abstract

The adoption of Artificial Intelligence (AI) in journalism and other communication practices brings up long-standing debates regarding the potentials of technological innovations for good and evil in society. Since the 20th century, when McLuhan argued that technologies help extend human capacity, media technologies have been regarded as liberating and empowering. Technologies aided human manipulation of mechanical and electronic processes in the media and communication industries. Arguably, social interactions were enhanced- extending audience reach, expanding scopes of coverage, altering the limitations of time and space, and bridging critical information gaps. By adding the power of computing to mechanical and electronic innovation of the past, as done with AI, far greater is the potential of media for good or ill in 21st-century society. The network societies are now better connected. Westernised societies are linked with those in the global south, individuals and media organisations alike are creating content. The resultant gluts of information further intensify the nature of global and social challenges. Given digital divide concerns being accelerated by AI, the Digital Dichotomy Theory (DD-Theory) is proposed towards understanding the inherent global media communication dynamics.

Keywords: artificial intelligence, dichotomy theory, digital, media and societies, machine learning

1. Introduction

Societies need information for many purposes in their journey to advancement. Whether it is for building the right physical infrastructure or for enhancing existing social structures, societies require the right kind of knowledge and information. As the central circulatory system carries oxygen to all parts of the body and expels the toxic substances, which could harm the body, the mass media are expected to infuse life-giving information to society, even the most remote members [1]. Access to required information helps dispel impediments on the path to the wellbeing of society, be this ignorance or adherence to discordant beliefs and thoughts. The media are expected to promote harmonious living in society. Technology was meant to enhance media efficiency.

Advancement in Digital Information Communication Technologies (D-ICTs) has heralded the arrival of Artificial Intelligence (AI). However, due to the digital divide across countries and continents, the gains are uneven across the world. As AI-based media communication imperatives are increasing potent aspects of knowledge-driven societies, there is an urgency to advance theoretical insights on the issue towards gaining a better perspective of media communication imperatives, especially about the position of a technologically dependent nation. Thus, this paper examines AI within the context of Dichotomy Theory to help interrogate the position of digital media communication dependencies. Premised on empirical inferences such as Technological Determinism as pre-existing theoretical frames, the paper argues that technologies may influence media communication imperatives in every society. However, there is a digital dichotomy and often affects the actual media communication outcomes, especially in developing countries like Nigeria.

2. Purpose and objectives

The paper aims to examine artificial intelligence and the media within the context of revisiting digital dichotomy theory. To achieve this aim, the following objectives are set:

1. To examine some of the perspectives on artificial intelligence.

2. To interrogate some conceptual framework on digital dichotomy.

3. To ascertain the realities of digital dichotomy amidst artificial intelligence.

3. Methodological approach

This paper is exploratory, as it utilises, the descriptive research method whereby relevant literature, documents, and records were consulted and analysed based on the subject matter. The paper is predominantly based on information derived from secondary sources, such as relevant texts, journals, official publications, historical documents, and the Internet, which served as tangible sources of insight into the analysis based on thematic areas. The method was used to evaluate such findings with other existing literature on the subject. The method help findings in the works available, check the consistency of such findings, and evaluates such findings with other findings.

4. Scoping review

The call by Nyam [2] to maximise the impact of AI in such countries, governments, and other stakeholders as well as communication scholars ought to put all resources and expertise towards meeting AI-oriented digital media communication needs of the society is adequate. Given digital divide concerns being accelerated by AI, the need to revisit the Digital Dichotomy Theory (DD-Theory) is important as this paper proposed it to be a better way of understanding the inherent global media communication dynamics. This is so because the basic assumption of the theory is that entities without the same predisposing factors will often significantly vary in the adoption time of current experience(s). Thus, AI does aid media communication realities to play out and affect humanity in such disparities.

Nyam [2] observed that the whole gamut of media classifications and applications, as well as operations, seem to be dependent on the available communication technologies. Today, digital media and communication had definitively advanced from basic software to AI. Sociology-Central in Nyam [2] affirms how the development of computers, for example, has increased audiences' spread and in turn made it more difficult to clearly distinguish between 'mass media' and 'non-mass media'. This expression relates to the contemporary influences of the new media upon the old 'traditional media.' The concept of 'new' applies to media technologies that have altered media classifications, with great contempt for communication characteristics of the traditional media.

Additionally, AI has advanced media communication reality. Notwithstanding, the regulatory framework is needed. The issue of the digital divide has indeed placed an extra burden on media scholars as well as professionals, and communication policymakers in developing countries. For instance, Adjei [3] mentions how old media, newspapers, television, and radio had the concept of feeding information based on the ground research for their listeners and viewers' in places such as Ghana, where radio and television stations tailor niche agenda-driven programs of political parties.

Pate [1] observed that where technology has been efficiently harnessed for the social, economic, and cultural wellbeing of groups and nations, a knowledge society emerges. Media technologies have always been a concern. Sometimes they had been viewed from the wide-angle lens of their facilitation of development communication goals, politics and good governance, the institution of democratic culture equality, and social justice. At other times, innovations in media technologies are viewed more narrowly within particular sectors, such as particular forms of messages, scope, and nature of communication enabled. The goals in health communication and marketing communication are likewise how to effect desired social behaviours. The concern in simplest terms is whether societies are never simple. As such, further questions are raised beneficial for which strata in society, under what conditions, and to what ends? These are the concerns evident in this paper regarding Artificial Intelligence.

Most of the African countries are broadly classified as developing. "As rapidly as technology is developing in the rest of the world, in Africa, things have moved at a slower pace," ([4], p. 52). The implication is that the global media imperative may have fundamental influences, but media experiences in developing nations are lagging. In this perspective, the position of the digital dichotomy is clear. The theory offers explanations to the power of media communication landscapes, and experiences between developed (invention driven media communication environment), and developing countries (adoption driven media communication environments). This has resulted in varying rates of AI-based digital updates and a 'global village.' Yes, this may be a global village, but the 'globe' has unequal media communication digits.

It is apparent in the literature that the adoption of Artificial Intelligence (AI) in journalism and other communication practices brings up long-standing debates regarding the potentials of technological innovations for good and evil in society. The paper, therefore, beamed the light on contemporary manifestations of global challenges, though understandably, the Nigerian context features prominently. Still, within the context of the literature, findings are shadowed by unprecedented global occurrences; the world has been bedevilled with a range of these in recent times. The paper validates the theoretical postulations that stark the double-edged sword that media technologies can be.

Arguably, since the 20th century, days when McLuhan argued that technologies help extend human capacity; media technologies have been regarded as liberating and empowering. Technologies aided human manipulation of mechanical and electronic processes in the media and communication industries. Similarly, social interactions were enhanced – extending audience reach, expanding scopes of coverage, altering the limitations of time and space, and bridging critical information gaps. With these came the potential to shift the balance of power in societies as desirable in democratic societies.

As observed by Pate [1] by adding the power of computing to mechanical and electronic innovation of the past, as done with Artificial Intelligence, far greater is the potential of media for good or ill in 21st-century society. The networked societies are now better connected. Westernised societies are linked with those in the global south, individuals and media organisations alike are creating content. The resultant gluts of information further intensify the nature of global and social challenges. The preceding arguments have created an important knowledge vacuum in the literature for this paper to be conceived.

5. Perspectives on artificial intelligence

Scholars like Wilson [5] observed that the rapid development of Artificial Intelligence (AI) heralds an era, one of machines or devices that are capable of learning by themselves (machine learning), and of imitating the human thoughts. The processes and concepts that relate to AI have been around since the 1950s. The term was coined by John McCarthy in 1955 and was popularised in 1956 at a research congregation in Dartmouth College in the United States. Furthermore, the United States Department of Defence focused on the development of Ai in the 1960s and produced computers to imitate basic human reasoning. Casey [6] remarks that although, AI is not new, it has become a technology of immense significance that anyone can hardly predict precisely where it is heading.

Within this context, AI is about systems that can learn and evolve through experience, which would most times carry our specialised tasks in gaming, decisions making and to transform large, complex, ambiguous information into real insights, to solve some of the world's most enduring problems. Sraders [7] sees AI as the science and engineering of making intelligent computerised machines that are programmed to closely imitate human thoughts and actions for the purpose of analysing data to address a variety of problems or execute tasks. It is a computer science filed that ensures the creation of intelligent computerised machines which are enabled to perform tasks, which normally requires human intelligence. These tasks include speech recognition, translation between languages, visual perception, etc.

Although AI is generally a broad term, there are different types or kids of AI, designed to perform different tasks. For example, there is specialised and general AI. Sraders [7] states that specialised AI is AI that is programmed to perform a specific task. Its programming is meant to be able to learn to perform a certain task – not multiple. On the other hand, general AI is not limited to one specific tasks- it is able to learn and complete numerous different tasks and functions. In general, much of the cutting-edge, boundary-pushing AI developments of recent years have been general AI.

AI is made up of a large variety of sub categories and areas in which they are applied some of these sub categories and the advanced abilities they offer include:

a. Machine learning: machine learning mimics human learning patterns, to gain an understanding of unstructured data sets and generate intelligent decisions such as medical decision making, Healthcare analytics, Bioinformatics, Emotional detection, Fraud detection, Cyber Security, Procurement optimization, Customer interactions and Optimised gaming.

b. Natural Language Processing (NLP): this permits an accurate analysis of data sets and communication of insights that touch on Communication systems, Legal assistants, Cognitive retail, Personal assistants and Web speech

c. Machine perception: simulate the human perception of the environment and extracts information from different data sources. For example, Medical imaging, Manufacturing, Service industry, Financial industry, Autonomous delivery, Transit safety, Geospatial analytics and Childcare.

d. Predictive analytics: analyse historical data to predict future outcomes. For examples, Marketing, Data extraction, Social Network analytics.

Therefore, in recent times, AI has risen to the forefront of public discourse because of its significant influence in the areas of cloud computing, big data, the Internet of Things (IOT), virtual reality and its potential to bring new possibilities for global development [8]. AI is already transforming web search, advertising, e-commerce, finance, logistics, media, and several other areas. The target of AI technology is to provide systems that would enable human-like interactions with software and provide decision-support for specific tasks [9].

Although AI technology is very effective for certain specific tasks, it is still limited and far from matching the highly diverse cognitive abilities of humans. There are still deficiencies in the AI technology. For example, virtual assistants such as Orange's Djingo, Amazon's Alexa, etc. cannot yet respond to questions using natural language, but this is surmountable in not too distant future. Re-echoed some of the limitations of AI to include data labeling, which has to be done by human, explainability problem, generalizability of learning and bias in data and algorithms, all of which would require human assistance for now.

6. Conceptual framework on digital dichotomy

Digital dichotomy simply refers to the digital divide. It is the centre of the conceptual frame of this paper. This hitherto referred to as 'technological divide.' As technologies have progressed into the digital phase, the divide has expanded more into a digital dimension-hence the term 'digital divide.' It has been the hallmark of persisting debate between developing nations and the otherwise developed ones. This is as a result of global media being a huge empire built on several years of inventions and innovations that have in turn been consistently improved upon. This technology remains dominated by the West (the large information-developed Northern hemisphere).

Therefore, Nyam [2] is of the view that many countries have at one point or the other lamented that the technical capacity of the Western media has been abused towards information flow disorder against developing nations. This position was largely termed the New World Information and Communication Order (NWICO) debate. In ensuring dynamics, the international media, many of which are based in Europe and North America, as well as modern Asia are believed to have the capacity to influence the media outcomes of developing societies, mainly in Africa and South America.

Within this context, scholars like Ozuru and Ekeanyanwu [10] remarked how communication at the international level comes with many consequences. Some of these consequences arise because of some imbalances, news manipulations, and sometimes, misrepresentation of some nations and people in the media systems of others. Corroborating this, Ciboh [11] observed that in 1973, governments of non-aligned nations met and discussed media and information flow issues, suggesting ways to counter the real or perceived imbalance.

Based on the preceding, there is an apparent digital dichotomy. The global digital divide is not denied, except there is a feeling that it is not a very valid point that can devalue the role of digital technology in much of modern existence. The global divide describes the unequal distribution of information, and communication technologies across nations. It has become a description for the information-have, and have-nots, although, much of these positions are complex to understand. In the words of argued that within academic circles it is well established that the digital divide encompasses more than physical access to D-ICTs. It is also a function of how D-ICTs are used. It is crucial to develop policies and programs that would bridge the global digital divide through D-ICTs.

For instance, former United Nations Secretary-General, Kofi Annan agrees that the digital divide is a serious issue, Annan's successor, Secretary-General Ban Ki-moon, admits, and leaders of the World Bank think so too. President James Wolfensohn, former World Bank even described the divide as "one of the greatest impediments to development." However, the significance of the digital divide has been challenged on several occasions, like Bill Gates thinking that the digital divide deserves no special attention because it is simply a symptom of economic disparity across nations, and thus the lack of access to information technologies in developing nations merely reflects the poverty level of those nations. Gates at a conference on the digital divide said "most of the world doesn't have cars, but we don't talk about the auto divide." Steve Jobs, Co-founder of Apple, reiterated the views saying that the so-called "digital divide" is "just a new sticker that people use to cover up a more important word: poverty."

In whatever point critics look at it, the significance of the digital divide becomes apt when culture, and media orientation of audience from a technology-adopting environment fail to key into the original intentions of inventors, as compared to audiences from a technology-inventive environment like the United States. Again, the digital divide becomes a more serious issue when the economic, and political policy, legal framework, and infrastructure of developing technology-adopting nations fail to meet up with international standards, and best media-communication practices.

The essence of digital technology is what prompted the conviction that the world is "truly" global. Yet some scholars are still sceptical that the export of digital technologies has not fully bridged the gap between developed, and developing worlds, because the hitherto less developed third world has not been able to conquer attraction to media contents of the West.

Digital technology is the reason for the conviction that the world is "truly" global. However, some scholars are still sceptical that the export of digital technologies have not fully bridged the gap between developed and developing worlds, because, the hitherto less developed third world have not been able to conquer attraction to media contents of the West. The West may also be genuinely interested in media contents of most developing countries, except that media production capacities and qualities have to be consistently upgraded even as the technologies may remain import-based [2]. See the manifestation in the diagram below:

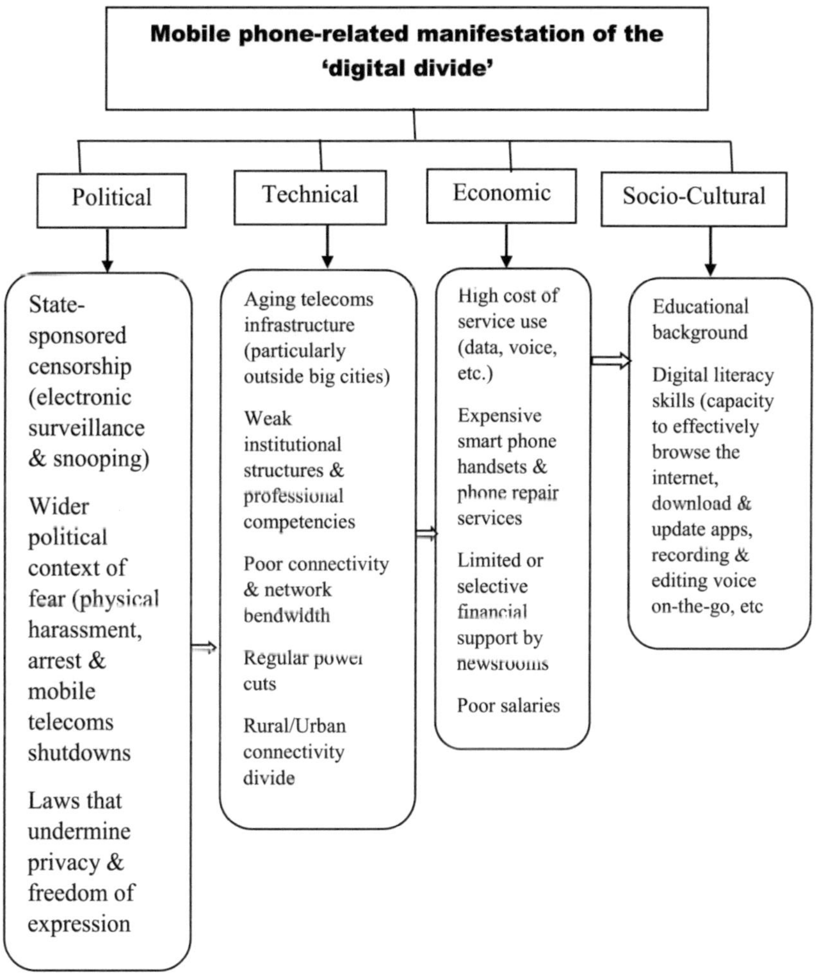

Source: (Mabweazara in [2]).

7. Theoretical framework

This paper finds the tenets of Technological Determinism, and Mediamorphosis Theories imperative. The two theories are considerably used in this paper together in a non-exclusive sense.

Technological Determinism applies to this paper because of its generic nature towards understanding how technologies are not just the base for mass communication, and contemporary mass media operations, but also how changes in technologies are determinants for changes in society, and respective media thereof. In other words, the theory applies to the generic influence of technology on humanity [12]. The Mediamorphosis Theory, on the other hand, is more specifically postulated towards a framework for understanding the constantly changing practices, and application in the media industry that can only be attributed to technology, which is hardly attributed to anything other than the technologies employed. As observed by Baran and Davis [13] Marshall McLuhan postulated the Technological Determinism Theory in 1970 towards predicting, and evaluating the role of all technologies. The explicit position relates to how technologies have been and are expected to transform media organisation, and experiences.

Thus, the two theories appear to be of the same continuum. While Technological Determinism is about the sociological implications of technologies in general, Mediamorphosis is particularly the implication of technologies to media convergences, and the opportunities for dynamic media orientations in the new, and conventional media [14].

Therefore, the adopted theories apply to this paper because of the importance of technology to society as well as the mass and the new media. Moreover, the role of technologies in the changing, constantly improving, but also diversifying forms of contemporary media, and communication means these theories are relevant. The basic assumptions, implications, and relative applications of information communication technologies justify this comparative analysis of the operational differences of the new, and traditional media, especially across societies at varying levels. For example, Sayad [15] corroborates that 60% of teachers across the world are not actively going to deal with D-ICT; just as besides 95% of students are not actively going to school, "digital mentoring" remains a key element for quality in education- lack of such digital aspects to education makes "10 points difference in learning within a country (micro-regions)."

8. The realities of digital dichotomy amidst artificial intelligence

In the case of the developing world, most of the advanced nations are fast employing legislation towards catching up with the uses, and applications of the new media amidst, or without synergies with the traditional media. Another flashpoint is in the area of investment. Governments and the corporate, or civil society in most developing countries are yet to call to question the urgency of digital technology, let alone understand the scientific cost that is involved over time. The advanced world plays hugely as they continue to enjoy and export to the digital developing countries. Satellite technology, for example, which tends to be dependent significantly on digitisation, is constantly being maintained and researched by the developed world [16].

Already, the Telecommunication Development Bureau (TDB) of the International Telecommunications Union is advocating for worldwide network relative understanding, and collaboration among policymakers, and regulators. Prefer to call "disruptive" or "destabilising" technologies. Others in the developed world seem to favour the term "transformative" technologies. Thus, technology is currently being deployed in almost every facet of our most recent civilisations, and modern life context. In this perspective, complex mobile networks such as

G5 are heralded along with increased technical and human operational intricacies. As such, the developing societies would need to catch up in terms of not just computational intelligence, but also perception intelligence, and cognitive intelligence.

Similarly, regarding the digital dichotomies, the adoption of ICT is seriously accelerating. The diffusion rate is rapid but also leaves more gaps and or consequences across societies with varying levels of development. As noted early, theoretical assumptions that enable sensible assumptions about contemporary media communication do exist. However, instances of proportional frame of reference to new media and communication such as Technological Determinism Theory are so far limited to understanding the spread, and influences of technology, and far less about what has, or can hinder or limit the overall benefits of D-ICTs. This is where DD-Theory fits in as a propositional frame of reference towards making improved technology and relevantly improved D-ICTs. Indeed, DD-Theory stands relevant as a new theoretical frame of reference for appraising the increasing global media-communication imperatives.

Besides, the status of technology in development is mainly accelerating and concentrated in developed wealthier nations, such as the United States, China, and European Union. New media realities in developing societies, such as media self learning, self-controlling, and self-communication stand-alone intelligent system [15] would demand rapidly improved understanding, or relative media-communication dichotomies across the world be enabled.

Entities without the same predisposing factors will often significantly vary in the adoption time of current experience(s). Adoption is not just due to capacity, but also time lapse-effect in the spread of invention orientation, and practice. This perhaps may be the reason why Ngwainmbi [17] concluded that a more limited form of globalisation might emerge just as there is a tendency for under-developed, and developing societies to over-depend on the so-called "world superpowers" for their protection. In line with the relative conclusion, Ngwainmbi [17] notes that the operational meaning of "superpower, advanced country", has to be redefined by scholars, political readers, media practitioners as well as knowledge-driven policymakers.

Nyam [2] cap it all, by saying that it is encouraging to collaborate towards improved global digital media-communication experiences. Aspect such as technological algorithmic innovations are needed at varying levels across nations, and journalism professionals, need to improve towards prevention or limiting hate speech, enhancement of fact-checking mechanisms, ethical encryption media practices among other merits. Irrespective of the ongoing advancements in network amidst digital dichotomy, such global D-ICTs conscious, and cautious collaboration can enable better learning among security operatives, digital rights literacy, and relative laws, as well as reasonable accountability from social media providers, and users.

9. Conclusion and recommendations

This paper examines the fundamental issue of digital dominance in information technologies. The paper interrogates how developing countries may, or have been left behind in the journey towards building knowledge network societies because of poor technological infrastructure, and systems. In particular, the paper examines the challenges relating to the communication, instrumentation, and monopolisation of network technologies, and the impacts of this on developing economies. This analysis rekindles the global information order of the past, such as media dominance, information inequity, asymmetrical, and imbalanced information flows. The paper proposes a new way of addressing the extant inequities and inequalities.

The paper adopted the position of Nyam [2] and postulates the Digital Dichotomy Theory (DD-Theory), that once there are significant differences in the predisposing factors of society, there will be digital technology adoption differences that would occur. Such difference will not be just due to financial, and physical capacity, but also due to time lapse-effect in the spread of invention, orientation, and practice(s).

Again, what makes a village? The world is not truly a "global village" as regards the dictum by McLuhan, and it will be difficult to be because there will always be a digital dichotomy between entities. There exist forms of a digital dichotomy because of the following reasons: the adoption difference(s) in previous technologies; dynamism in cultural, economic, political, and religious systems of entities across the globe; the time and space lapse between invention(s) entities, and adoption entities. Mere resistance to change, change cannot be forced but persuaded.

There is a digital dichotomy that places developing societies on the side of playing catch-up, governments, and citizens must be aware, and active in the ongoing digital technological imperatives. Besides, governments in many nations still hinder, and or censor global, and local information. AI may be taking undue advantage of such unfortunate dynamism of improved digital communication (Jimoh in [2]). He argued that this is not about the future of media communication in developing nations or states, but for the overall advantageous possibilities, and convergences of the 21st century.

The paper subscribes to the call of Nyam [2] that in a global media scenario, developing societies cannot afford to significantly lag. It is good that developing countries with huge human and natural resources should be challenged to be on the information superhighway. This may serve better than otherwise. Also, this is expected to harvest more towards development. However, research, and training in media professionalism, and computing (programming, hard or software engineering, internet security, among others) are strongly recommended towards maximisation of the convergences, and synergies of media forms. Digital technologies depend on excellent software programming and networking.

Author details

Aondover Eric Msughter[1*], Aondover Ogechi Perpetua[2] and Adebija Lillian Itiafa[3]

1 Department of Mass Communication, Skyline University, Nigeria

2 Faculty of Communication and Media Studies, Department of Cinematography and Broadcast Studies, Rivers State University, Port Harcourt, Rivers State, Nigeria

3 Department of Mass Communication, Bayero University, Kano, Nigeria

*Address all correspondence to: aondover7@gmail.com

References

[1] Pate AU. Knowledge societies: Artificial intelligence and the media. In: Fayoyin A, Ademosu I, editors. Foreword. UNESCO West Africa Regional Office: UNESCO; 2021

[2] Nyam II. Knowledge societies: Artificial intelligence and the media. In: Fayoyin A, Ademosu I, editors. Digital Dichotomy Theory: Towards Propositional Appraisal of Artificial Intelligence Based Media-Communication Imperative. UNESCO West Africa Regional Office: UNESCO; 2021

[3] Adjei E. The media, artificial intelligence, Nation States and the future. A Paper Presented at the 1st virtual 7th Annual Conference of the Association of Communication Scholars & Professionals of Nigeria. UNESCO West Africa Regional Office; 2020

[4] Smith JA. Interpretative Phenomenological Analysis: Theory, Method and Research. New York: Sage Publication; 2009

[5] Wilson J. Knowledge Societies: Artificial Intelligence and the Media. The Fake News Phenomenon: Can AI Technology Stem the Tide in Nigeria? UNESCO West Africa Regional Office: UNESCO; 2021

[6] Casey K. 5 AI fears and how to address them. 2019. Available from: https://enterprisersproject.com/article/2019/9/artificial-intelligence-ai-fears-how-address

[7] Sraders A. What is artificial intelligence? Examples and news in 2019. 2019. Available from: https://www.thestreet.com/technology/what-is-artificial-intelligence-14822076

[8] UNESCO. Artificial Intelligence for Sustainable Development Synthesis Report: Mobile Learning Week 2019. Paris: UNESCO; 2019

[9] Wilson J, Jibrin R. Mitigating hate speech in Nigeria: The possibilities of artificial intelligence. Analisi: Quadarns Comunicacio Cultura. 2019;**6**(1):17-30

[10] Ozuru E, Ekeanyanwu N. Audience assessment of the influence of social media network on the global news flow controversy. In: Wilson D, editor. Communication and the News Media in Nigeria Social Engagements, Political Development and Public Discourse. Nigeria: African Council for Communication Education; 2013

[11] Ciboh R. Global Media and Culture Domination. Pacific University: CherryBrooks; 2005

[12] Nwuneli O. Communication technologies and globalized identities. A Keynote Address Presented at 1st virtual and 7th Annual Conference of the Association of Communication Schoars & Professionals of Nigeria (ACSPN), in Collaboration with UNESCO, IFAD & COMEST. September 2-3. 2020

[13] Baran S, Davis D. Mass Communication Theory: Foundations, Ferment and Future. Wadsworth Publishing Company; 2010

[14] Anaeto S, Onabaajo O, Osifeso J. Models and Theories of Communication. University of Lagos: Arica Renaissance Inc.; 2008

[15] Sayad A. Education for the future, means respond to the present. A Paper Presented at the 1st Virtual and 7th Annual Conference of the Association of Communication Scholars & Professionals of Nigeria (ACSPN). UK, London:

Collaboration with UNESCO, IFAD & COMEST; 2-3 Sept 2020

[16] BBC. How Worried Is Silicon Valley about Safe Harbour? BBC news; 2015

[17] Ngwainmbi E. Mystifying and demystifying shared information on Covid-19 across cyber mediated news networks: A scioliogical approach to solutions for vulnerable populations. A Paper Presented at the 1st Virtual and 7th Annual Conference of the Association of Communication Scholars & Professionals of Nigeria (ACSPN), in collaboration with UNESCO, IFAD & COMEST. September 2-3. 2020

Chapter 9

Hybrid Perovskite-Based Memristor Devices

Mansi Patel, Jeny Gosai, Nitin Chaudhari and Ankur Solanki

Abstract

Modern electronic devices are being developed for cutting-edge applications, as a result of recent developments in artificial intelligence (AI) and machine learning (ML). The demand for "universal memory" devices with exceptional qualities, such as high data transmission speed, storage capacity, non-volatility, and low operation voltage has increased as a result of the industry's ability to sustain such a high growth rate. In this chapter, we elaborate on the history of the evaluation of novel memristor structures, various switching mechanisms, and materials for developing memristor devices. The current state of the art of the memristor for various applications, such as data storage, artificial synapse, light-induced resistive switching, logic gates, and mimicking human behavior is also systematically summarized.

Keywords: hybrid perovskite, data storage, neuromorphic computation, resistive switching, memristor devices

1. Introduction

Since the discovery of the first programmable computer, the dependency associated with paper and canvas-based mediums to store and process information has significantly declined. According to Moore's law and Dennard's law, the technology based on the traditional complementary metal-oxide semiconductor (CMOS) has great strides in the last few decades and led to a sharp rise in digital capabilities [1–3]. In traditional computing technology known as Von Neumann architecture, the system comprises two separate units, namely central processing unit (CPU) and data storage unit, connected through the bridge known as data busses. This traditional architecture causes the delay in processing and consumption of more power in the process, introduced as bottleneck of Von Neumann architecture. The memristor has emerged as a novel device to improve or develop new technology based on the fusion of memory and processor. In the last few years, memristor has captured the significant attention of researchers due to its excellent properties, such as simple structure, high-density data storage, low power consumption, fast switching speed, long endurance and retention, multistage and high scalability. Due to these properties, memristors can be used for artificial intelligence (AI), the Internet of Things (IoT), wearable electronics, smart medical applications, logic circuits, neuromorphic computing, etc. [4, 5].

In this chapter, we discuss the history and various switching mechanisms in memristors. Different including organics, inorganics, and hybrid materials have been discussed to use as active layers for memristive applications. Metal halide organic–inorganic perovskites are well taken as an example of hybrid materials due to outstanding electrical, optical, and structural properties. Various applications of memristors, such as data storage, logic gates, and photonic devices, including many bionic electronic systems as artificial synapses, neural networking, nociceptors, artificial retina, etc., have been summarized.

2. History of memristor

Before the discovery of the memristor, resistor (1827), capacitor (1745), and inductor (1831) were considered only three fundamental passive circuit elements. In the year 1971, Leon Chua theoretically proposed the fourth fundamental element named memristor (memory resistor), which shows the relation between charge and flux. Few decades later, in 2008, a strong connection between Chua's theory and the experimental model was observed in Hewlett Packard lab and the first physical memristor model based on TiO_2 material was realized [6]. This prototype memristor showed the data storage capabilities, processing logical operation with long retention time and low operating voltage, as a result of the change in their resistance states [6, 7]. Due to its potential scalability and low power consumption for memory applications, memristor continues to stimulate a steady expansion in the research industry on a global scale [8].

Memristor is a non-volatile two-terminal electrical component with a sandwich structure called metal–insulator–metal (MIM), as shown in **Figure 1(a)**. Memristor is basically dependent on charge and magnetic flux also called memristance, which varies as a function of the electric charge (q) and magnetic flux (φ) (**Figure 1(b)**). This property cannot be obtained by any relation of the other fundamental elements, such as resistor, capacitor, and inductor [9, 10]. A most curious feature of the memristor is its memory function, which originates from its resistance states [10]. Memristors are devices that switch between low resistance state (LRS) and high resistance states (HRS), according to applying voltage bias, by applying positive bias memristor resistance changed from HRS to LRS at a particular voltage (set voltage or VSET) and opposite applying negative bias resistance state changed from LRS to HRS (reset voltage or VRESET). This feature may be used to store data in a resistant state with high adoption properties [11].

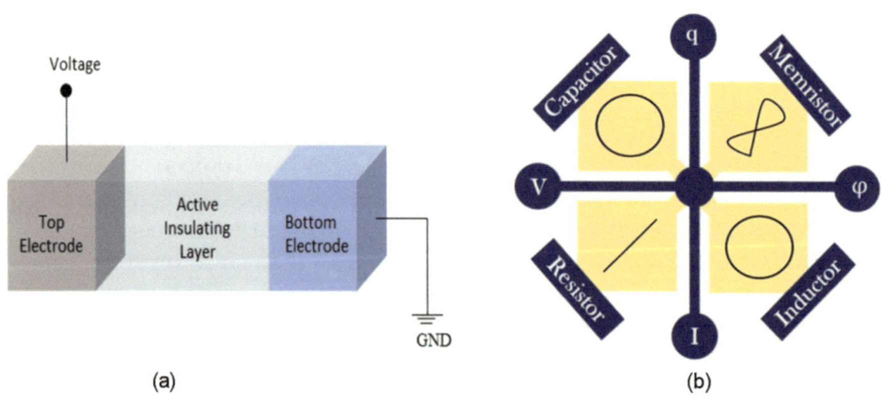

Figure 1.
Illustrations of the (a) MIM structure of memristor (b) I–V characteristics of the fundamental circuit elements.

3. Classification of Memristors

Based on the switching mechanism, memristors have been classified into three different types: resistive switching, ferroelectric, and phase change (as shown in **Figure 2**).

3.1 Resistive switching (RS)

Resistive switching is the one of most prevalent types of memristor devices that was first observed in TiO_2-based devices. The RS mechanism is one of the most complicated and also conflicts between number of parameters, such as electrodes/active layer interfaces, grain size, active area of the device, types of defects, and many more [12]. The basic characteristics of resistive switching are based on the movement of ions, such as oxygen vacancies [13–15], active metal cations [3, 16, 17], and anions, like halides [18, 19] and sulfurs [20]. The two most common type of resistive switching mechanism is the electrochemical mechanism (ECM) and valence charge mechanism (VCM) as discussed below:

3.1.1 Electrochemical mechanism (ECM)

Here, an electrochemical redox reaction, at the active electrode under an external voltage applied, produces the RS characteristics. Initially, a positive voltage was applied to the active electrode (top electrode Ag), which caused the metal atoms (Ag) to oxidize and transform into the corresponding ions (Ag^+), which can subsequently diffuse through the active layer to the bottom inert electrode and reduced to Ag atoms at the interface. There are two resistance states in memristor: low resistance state (LRS) and high resistance state (HRS), which refers to the formation and rapture of the conduction filaments in the active layer between the top (Ag) and bottom electrode, respectively. This reaction represents a common electrochemical oxidation–reduction process. Some other materials used as active electrodes are Cu, Au, Pt, and Al [21, 22].

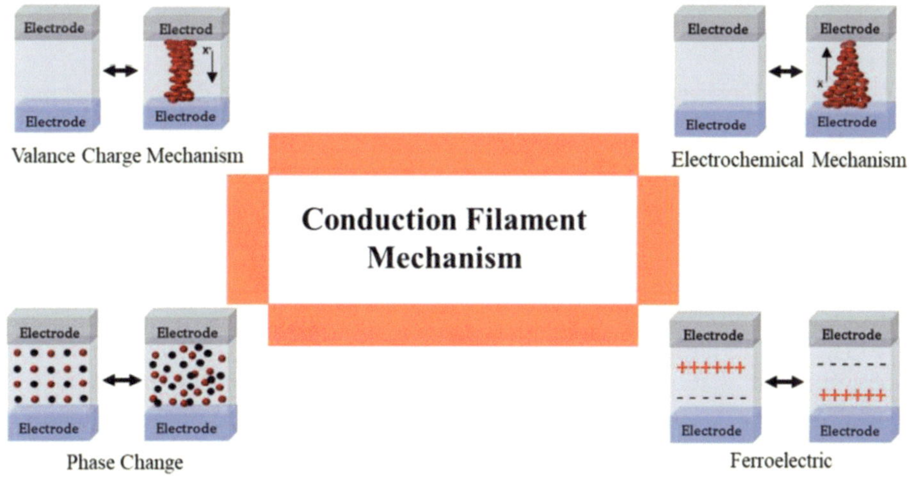

Figure 2.
Illustration of types of switching mechanisms in various memristors.

3.1.2 Valence charge mechanism (VCM)

In terms of integration and scaling, filamentary VCMs are the most sophisticated mechanism. The two electrodes are shorted by the formation and bursting of conductive filaments (CF), which are caused by a concentrated localized area of defects. There may be two or more stable resistance states, depending on how the CF diameter and/or dissolution are modulated or controlled. The conductance of interfacial VCM devices is assumed to scale with the device junction area through a homogenous oxygen ion flow across the oxides, either at the electrode/oxide or oxide/oxide interface. Bilayer stacks of difficult oxides, such as TiO_2/TaO_2 [23] and a-Si/TiO2 or complex oxides as bismuth ferrite [24] and praseodymium calcium manganite, form the foundation of reference material systems [25].

3.2 Ferroelectric

Despite the fact that ferroelectric memory has been around for a while, it has not captured the attention due to various scale-up issues. In 2006, the first device as ferroelectric tunnel junctions (FTJs) [26] was realized and led as novel concept for data storage and neuromorphic computing [27, 28]. An FTJ comprises two metal electrodes separated by a thin ferroelectric insulator. Quantum electron tunneling, in which electrons pass through a potential barrier of the ultrathin insulator, is the dominating process in this ferroic nanostructure, which is made up of the ultrathin ferroelectric barrier. The alignment of ferroelectric polarization in the insulator can change the stream of electrons, producing the enormous tunnel electro-resistance (TER) effect. Depending on the polarity of the ferroelectric layer, the tunneling electrons are either attracted or repulsive. An energy band profile becomes asymmetric when two distinct metals are placed across the ultrathin ferroelectric layer. When voltages are applied across the device, the electric potential at the interface increases or decreases depending on the polarization direction. As a result, the modified energy potential control the electrons transport through these contacts [29]. This screening phenomena causes the FTJ to exhibit significant resistance changes, allowing for the storage and processing of data [30].

3.3 Phase change

A device can be referred to as a hysteretic memristor when the joule heating led the phase changes between two states: amorphous and crystalline. Among all new memristor technologies, the phase-change-based memristor is the most developed and commercialized in the storage class memory (SCM) sector. One could say that phase-change memory technology has made a significant contribution to the growth of new electronic technologies. The first time, phase-change technique originally described was nearly 50 years ago [31]. However, recently, the phase-change technology has only gained popularity as a result of research on chalcogenide materials such as $Ge_2Sb_2Te_5$ [32] or Ag- and In-doped Sb_2Te [33]. **Figure 2** illustrates a phase-change memory instance [34]. The narrow metal heater's current saturation promotes the joule heating process when an electrical field is applied between the top electrode and bottom metal heater. An internal temperature change gradually heats the phase-change material [35]. The resistance contrast results from the phasechange between an HRS that is amorphous and an LRS that is crystalline. The distribution of internal voids determines the variation in structural disorder between two states. The band structure is redesigned as a result of ordered vacancies, and the conductivity rises as a result of the localization of charge carriers in a crystalline state [36].

4. Active materials for Memristor

A plethora of materials used as active layers for memristors can be categorized as biomaterials, organic, inorganic, and hybrid materials as discussed below:

4.1 Inorganic materials

Inorganic materials consist of metal oxides, halides, chalcogenides, 2D materials, etc. The main advantage of inorganic materials is their strong environmental stability in varied conditions. The first ever memristor was prepared by Hewlett-Packard (HP), composed of two inorganic layers: stoichiometric titanium oxide and oxygen-deficient non-stoichiometric TiO_{2-x} layers [6]. The conductivity was achieved by the migration of the oxygen vacancies when the external electric field is applied. Later, many other inorganic oxides, such as HfO_2, CuO [37], NiO [38], and TiO_2 [39]. attracted the attention of researchers. Many 2D materials were also used as the active layer, such as nitrides, transition-metal dichalcogenides (MoS_2 and WS_2) [40], InSe [41], black phosphorus (BP) [42], MXenes [43], bismuthene [44], and tellurene [45]. Few nanorod structures explored consist of ZnO [46], TiO_2 [47], HfO_2 [48], etc. **Table 1** summarizes some of the advancements in the materials used till date (**Figure 3**).

4.2 Organics materials

Organic active layer materials are generally small organic molecules, polymers (synthetic and natural), etc. These materials have gained a great deal of interest from researchers for resistive switching due to their attractive properties compared to

Material type	Material Name	ON/OFF ratio	Vset/Vreset (Voltage)	Endurance (Cycle)	Retention (Seconds)	Reference
Inorganic	Cuo	10	-1 V/3.2 V	100	2 x 10^4s	[37]
	Nitride	100	-1.5 V/2 V	—	—	[19]
	Black phosphorus (BP)	2 x 10^7	-2 V/1 V	100	10^4	[42]
Organic	P3HT	10	—	1000		[50]
	PEDOT:PSS	10	4 V/-3 V	1000		[51]
	PCBM	400	0.5V/-1V	47	—	[52]
	PFT-PI	10^4	-2 V/2 V	180	10^4	[53]
	Cu-TCNQ	10^3	3.5 V/0.4 V	10^3	—	[54]
	PVK	10^4	-1 V/3.3 V	—	10^4	[55]
Hybrid	$MAPbI_3$	10^2	0.7 V/−0.61 V	600	10^4	[56]
	$CsPbI_3$	10^5	−0.95 V/−0.71 V	100	10^4	[57]
	$(PEA)_2MA_4Pb_5I_{16}$ (quasi-2D)	10^4	0.15 V/-1 V	500	300	[58]
	$(IFA)_3PbI_5$ (1D)	10^3	0.2 V/−2.1 V	200	10^4	[59]
	$Cs_3Sb_2I_9$ (0D)	10^2	1 V/-1 V	500	5000	[60]

Table 1.
Summary of reported materials for memristors.

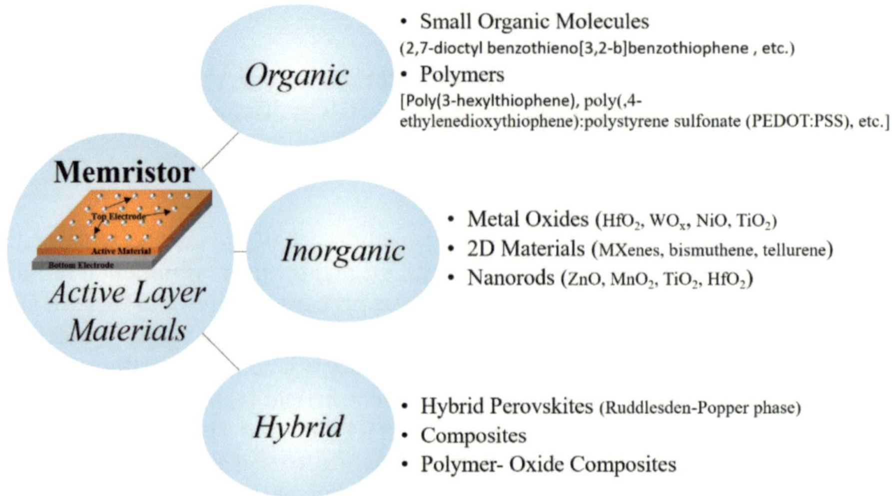

Figure 3.
Illustrations of types of active materials used in memristors.

traditional metal oxides [61]. The properties include easy solution processing and good chemical, mechanical and morphological properties; also, it possesses high intrinsic flexibility. Although some of the organic molecules lack in stability portion; they overcome it by having a low operating voltage as well as a good memory window. These materials can serve as a good candidate for low-power operating memory systems. Some materials, such as poly(3-hexylthiophene) (P3HT) [50], poly(,4-ethylenedioxythiophene): polystyrene sulfonate (PEDOT:PSS) [51], phenyl-C61-butyric acid methyl ester (PCBM) [62], phenanthrol [9,10-d] imidazole (PFT– PI) [53, 63], cobalt(III)-containing conjugated and nonconjugated polymers [63], polyaniline (PANI) [64], copper-tetracyanoquinodimethane (Cu-TCNQ) [54], etc., have shown good performance not only as the active layer but also as electron transfer layer in transistors, light-emitting diodes, and photovoltaic devices. These materials usually have shown the dual filament formation as an RS mechanism including the phase change, redox reactions, conformation change, and charge transfer mechanism. Due to the advantages of small organic molecules, such as easy modification in electronic properties and easy processing, these molecules can be used as an active medium in synaptic as well as memory devices.

4.3 Hybrid materials

Hybrid materials are combinations of organic and inorganic components, for instance, graphene oxides and composites, polymer-oxide composites, hybrid perovskites, etc. The organic and inorganic components, when merged, can create some desirable property material with enhanced quality and diminished defects. These materials are quite popular for non-volatile memory applications having low power consumption and high speed. The first hybrid perovskite (MA)PbX_3 (MA = methylammonium and X = Cl, Br, I) was first reported by Dieter Weber in 1978 [65]. Perovskites have a general chemical formula ABX_3, where A occupies eight corners of the cubic unit cell, it is a large monovalent cation, such as methylammonium (MA^+) and formamidium (FA^+). B is represented by a divalent metal cation, occupying body central position in the cubic unit, for example, Pb^{2+}, Sn^{2+}, Eu^{2+}, Cu^{2+}, etc. X is the halide anion, such as I^-, Br^-, and Cl^-, six of which surround the

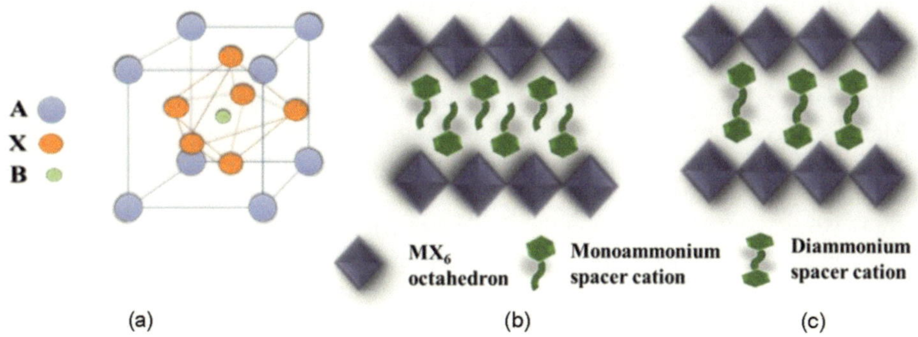

Figure 4.
Illustration of basic structure of (a) perovskite, (b) Ruddlesden-Propper phase, and (c) Dion-Jacobson phase.

B cation in the octahedral geometry. Making $[BX_6]^{4-}$ octahedron [66]. The structure of perovskite is depicted in **Figure 4**.

Hybrid organic–inorganic perovskites (HOIPs) are considered a recent family in the perovskites class. Due to the number of possible combinations of the A, B, and X, there are many perovskite family members reported, also they can form 3D, 2D, 1D, and 0D structures as well as having the same unit cell. The structural and chemical diversity offered by HOIPs give rise to tuning to achieve desirable properties and opening doors to many potential applications [67]. HOIPs have certain exclusive properties, such as tunable band gap, wide range of light absorption, ambipolar charge transport, long electron–hole diffusion length, and optical absorption, making them apt for devices such as solar cells, light emitting diodes (LEDs), transistors, and memristors. In 2016, the first time reported the resistive switching phenomenon and synaptic properties in three-dimensional (3D) HOIP ($MAPbX_3$, X = Br and I) devices, due to the presence of ion migration. HOIPs containing organic-based cations are hydrophilic in nature, thermally unstable, and immediately degrade in ambient air. Furthermore, inorganic materials are mixed with organic materials to reduce operating voltage. By decreasing dimensionality from three dimension (3D) to two dimension (2D), increasing quantity of organic insulating cations, which increase the activation energy of halide ions in perovskite layer. As a result, 2D has a high amount of insulating cations with the ability of lowest magnitude of energy consumption. Mainly, memristors operated due to the ion migration has reported, but several studies also claim that metallic filament growth plays an important role in resistive switching properties. For synthesizing stable materials, low-dimensional perovskite phases, namely defined as Ruddlesden-Propper (RP) and Dion-Jacobson (DJ) perovskites, which distinguished by the change in their interlayer spacer cation alignment (**Figure 4b** and **c**) [4, 68, 69].

5. Applications

Perovskite materials have the potential to change properties with light; because of this property, this material is used in the application of optoelectronic as well as in data storage devices. By combining both to prepare optoelectronic based logic gate devices, and this electrical signal is used as write, and the optical signal is used as erase in resistive switching devices and by changing the light intensity increase/decrease set and reset voltage [70, 71]. Wang and colleagues programmed the Au/$MAPbI_{3-x}Cl_x$ structure devices to set/reset by photo/electrical bias [72]. For multilayer storage RRAMs, the set voltage falls as light intensity rises. The gadget can carry out logical operations and coincidental event detection tasks by using optical and electrical pulses. In their experiment, Chai and colleagues discovered that light might lower the device's set voltage, and this finding might be used to develop logic circuits [73].

Memristors can potentially be useful in ultrahigh storage density computing technologies. An immediate application for these devices is the resistive random-access memory (ReRAM). To meet the growing demands of next-generation data storage devices, ReRAMs must exhibit characteristics such as small write voltage (few hundred millivolts), short write time (<10 ns), small read voltage so that there is no change of internal resistance, high OFF-to-ON resistance ratio (>10), high endurance ($\sim 10^3$), high retention (~10 years), and small device size (<10 nm), in addition to low-cost fabrication and flexibility [74].

Memristor can be used as a programmable logic gate with the building of crossbar architecture. When comparing CMOS-based devices with memristor devices are far more variable. The major problem in many logic gate architectures using memristor is the endurance and device-to-device variation. Also, memristor-based gates become less reliable but the capacity to accept changes in weight values has a high level of device variability tolerance.

Memristors are further used in artificial neural networks. Here, two of them are discussed, CNN (convolution neural network) and SNN (spiking neural network). **CNN (convolutional neural network)**: Memristor arrays allow for the concurrent and as well as on computation of vector–matrix multiplication operations, which significantly speeds up inference and training for convolutional neural networks (CNNs) and related deep neural networks (DNNs). Through Kirchhoff's current law and Ohm's law, the memristor crossbar arrays are employed to store the weights and carry out simultaneous multiply-accumulate operations [75]. In this system, the outputs are represented as the accumulated currents on the columns, while the inputs are represented as voltage pulses applied to the rows. ADCs or a sense amplifier may then read out the output activations after being quantized. First single layer perceptron is used to demonstrate simple pattern recognition, which serve as the foundation for memristor-based artificial neural network development [76, 77]. Later these work on multi-layer and CNN based architecture used for image recognition from well-known data sets as MNIST and CIFAR 10 [78]. **SNN** (spiking neural network): Memristors' internal dynamics and processes bear strong resemblances to biological processes, opening the door to the development of bio-faithful neuromorphic systems without the need for intricate circuitry and sophisticated algorithms. Information processing can instead be carried out locally by device dynamics. SNN is one of the applications of the internal dynamics based on memristor devices, this system maximizes the efficiency of complex functions. SNNs have the potential to mimic the biological brain closely due to the owing spike-driven communication. This system fires a signal when the potential from input reach to threshold and the fired signal transfer to the neighbor neuron. The STDP learning rule for SNN training comes from neuroscience. For train to SNN, two types of learning are used: supervised learning and unsupervised learning. The unsupervised STDP learning is very energy efficient because of the local learning nature and ability to learn in only few spikes. Where accuracy is considered the most important factor, supervised learning is generally utilized for training neural networks. The inability to differentiate spiking events is the main obstacle to backpropagation in SNN [79].

Memristor can also be useful as artificial vision sensor, touch sensor, and pain sensor in the arena of bionic electronics. Vision is the crucial sense system through which

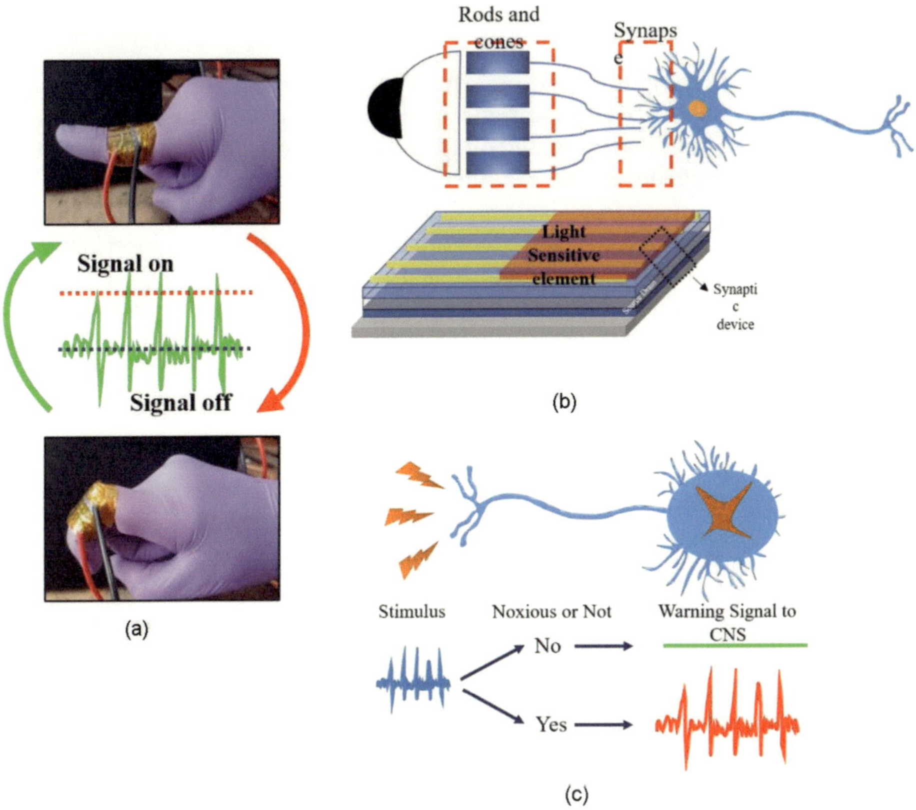

Figure 5.
Illustrations of the memristor-based bionic systems (a) memristor as a touch sensor, (b) artificial retina, and (c) nociceptor.

most of the information are collected by humans. To replicate artificial vision sensor, memristor is used as visual neuromorphic device. This involves two types of impulse signals, optical and electrical, as presynaptic and postsynaptic impulses as input and output signals, respectively. This kind of device is also called photonic memristor and that used as photodetector with photo stimuli as retain of human eyes that collect process visual picture and transport [80].

Another application is pressure or touch sensor, where a small electrical signal obtains as a result of pressure, which is promptly transferred to a sensory receptor as an input response for further processing. This system is only initiated when the pressure touch is converted into electrical signal and therefore postsynaptic current. As shown in **Figure 5(a)**, changing pressure on device by finger folding fire signal in form of electrical current. It can measure the quantity, frequency and speed, and time duration of that touch pressure [81].

In reaction to harmful stimuli, the body feels the unpleasant emotion of pain. When external stimuli activate pain receptors in the stomach or body, the central nervous system receives, interprets, and sends the pain information. The ability of AI systems to perceive pain and become sensitized to it is essential for significantly increasing the efficacy of hardware devices since it enables them to have different sensitivities to external stimuli for various purposes [82].

6. Conclusions

Memristor-based computing platforms have been proposed to give great processing efficiency for data-centric applications, in order to address the drawbacks of conventional computing systems based on the von Neumann architecture. Despite recent major advancements in this area, there are still a lot of obstacles to overcome. To address the stochasticity and CMOS compatibility difficulties, which have an impact on the underlying electrical performance, material system engineering is essential. The integration and reliability problems might have a potential answer to tackle. To solve the concerns with variability and device non-ideality, more material and device development is still necessary. However, novel materials and architecture, or circuits have not yet reached unraveling full potential of memristor for advanced technologies.

Acknowledgements

A.S. and M.S. gratefully acknowledge the financial support from Science and Engineering Research Board (SERB) core research grant CRG/2020/000869. A.S. would also like to gratefully acknowledge grant GUJCOST/STI/2021-2022/3873 from the Government of Gujarat, India, and ORSP/R&D/PDPU/2019/AS00/RO047 from Pandit Deendayal Energy University (PDEU) to perform this study. J.D. and N.C. would like to acknowledge PDEU Start-up grant ORSP/R&D/PDPU/2021/NC00/R0069.

Author details

Mansi Patel[1,2], Jeny Gosai[2,3], Nitin Chaudhari[3] and Ankur Solanki[1,2]*

1 Department of Physics, School of Energy Technology, Pandit Deendayal Energy University, Gandhinagar, India

2 Flextronics Lab, Pandit Deendayal Energy University, Gandhinagar, India

3 Advanced Hybrid Nanomaterial Laboratory, Department of Chemistry, School of Energy Technology, Pandit Deendayal Energy University, Gandhinagar, Gujarat, India

*Address all correspondence to: ankur.solanki@sot.pdpu.ac.in

References

[1] Moore GE. Cramming more components onto integrated circuits, Reprinted from Electronics, volume 38, number 8, April 19, 1965, pp.114 ff. IEEE Solid-State Circuits Society Newsletter. 2006;**11**:33

[2] Kawahara A, Azuma R, Ikeda Y, Kawai K, Katoh Y, Hayakawa Y, et al. An 8 Mb multi-layered cross-point ReRAM macro with 443 MB/s write throughput. IEEE Journal of Solid-State Circuits. 2013;**48**:178

[3] Choi J, Le QV, Hong K, Moon CW, Han JS, Kwon KC, et al. Enhanced endurance organolead halide perovskite resistive switching memories operable under an extremely low bending radius. ACS Applied Material Interfaces. 2017;**9**:30764

[4] Gogoi HJ, Bajpai K, Mallajosyula AT, Solanki A. Advances in flexible memristors with hybrid perovskites. Journal of Physical Chemistry Letters. 2021;**12**:8798

[5] Perrissin N, Gregoire G, Lequeux S, Tillie L, Strelkov N, Auffret S, et al. Perpendicular shape anisotropy spin transfer torque magnetic random-access memory: towards sub-10 nm devices. Journal of Physics D. 2019;**52**:234001

[6] Strukov DB, Snider GS, Stewart DR, Williams RS. The missing memristor found. Nature. 2008;**453**:80

[7] Yang JJ, Strukov DB, Stewart DR. Memristive devices for computing. Nature Nanotechnology. 2013;**8**:13

[8] Waser R, Aono M. Nanoionics-based resistive switching memories. Nature Materials. 2007;**6**:833

[9] Chua L. Resistance switching memories are memristors. Applied Physics A: Materials Science & Processing. 2011;**102**:765

[10] Chua LO, Sung Mo K. Memristive devices and systems. Proceedings of the IEEE. 1976;**64**:209

[11] Kumbhar D, Tripathi B, Patel M, Dongale T, Solanki A. Mathematical investigation of the hysteresis mechanism in hybrid perovskite-based heterojunction devices for the resistive switching applications. In: Presented at 2022 2nd Asian Conference on Innovation in Technology (ASIANCON). 26-28 Aug 2022

[12] Liao K, Lei P, Tu M, Luo S, Jiang T, Jie W, et al. Memristor based on inorganic and organic two-dimensional materials: mechanisms, performance, and synaptic applications. ACS Applied Material Interfaces. 2021;**13**:32606

[13] Yoon JH, Song SJ, Yoo I-H, Seok JY, Yoon KJ, Kwon DE, et al. Highly uniform, electroforming-free, and self-rectifying resistive memory in the Pt/Ta2O5/HfO2-x/TiN structure. Advanced Functional Materials. 2014;**24**:5086

[14] Yoon JH, Kim KM, Song SJ, Seok JY, Yoon KJ, Kwon DE, et al. Pt/ta2o5/hfo2−x/ti resistive switching memory competing with multilevel NAND flash. Advanced Materials. 2015;**27**:3811

[15] Jiang H, Han L, Lin P, Wang Z, Jang MH, Wu Q, et al. Sub-10 nm ta channel responsible for superior performance of a HfO2 memristor. Scientific Reports. 2016;**6**:28525

[16] Han JS, Le QV, Choi J, Hong K, Moon CW, Kim TL, et al. Air-stable

cesium lead iodide perovskite for ultra-low operating voltage resistive switching. Advanced Functional Materials. 2018;**28**:1705783

[17] Seo J-Y, Choi J, Kim H-S, Kim J, Yang J-M, Cuhadar C, et al. Wafer-scale reliable switching memory based on 2-dimensional layered organic–inorganic halide perovskite. Nanoscale. 2017;**9**:15278

[18] Han JS, Le QV, Choi J, Kim H, Kim SG, Hong K, et al. Lead-free all-inorganic cesium tin iodide perovskite for filamentary and interface-type resistive switching toward environment-friendly and temperature-tolerant nonvolatile memories. ACS Applied Material Interfaces. 2019;**11**:8155

[19] Kumbhar D, Jain M, Solanki A. Forming free non-volatile resistive switching mechanism in ruddlesden popper perovskite memristors. In: Presented at 2022 International Conference for Advancement in Technology (ICONAT). 21-22 Jan 2022

[20] Zhang P, Gao C, Xu B, Qi L, Jiang C, Gao M, et al. Structural phase transition effect on resistive switching behavior of mos2-polyvinylpyrrolidone nanocomposites films for flexible memory devices. Small. 2016;**12**:2077

[21] Saylan S, Aldosari HM, Humood K, Abi Jaoude M, Ravaux F, Mohammad B. Effects of top electrode material in hafnium-oxide-based memristive systems on highly-doped Si. Scientific Reports. 2020;**10**:19541

[22] Praveen P, Rose TP, Saji KJ. Top electrode dependent resistive switching in M/ZnO/ITO memristors, M = Al, ITO, Cu, and Au. Microelectronics Journal. 2022;**121**:105388

[23] Gao L, Wang IT, Chen PY, Vrudhula S, Seo JS, Cao Y, et al. Fully parallel write/read in resistive synaptic array for accelerating on-chip learning. Nanotechnology. 2015;**26**:455204

[24] You T, Ou X, Niu G, Bärwolf F, Li G, Du N, et al. Engineering interface-type resistive switching in BiFeO3 thin film switches by Ti implantation of bottom electrodes. Scientific Reports. 2015;**5**:18623

[25] Alessandro F, Severin S, Junwoo J, Pritish N, Robert MS, Geoffrey WB, et al. Bidirectional non-filamentary RRAM as an analog neuromorphic synapse, part I: Al/Mo/Pr0.7Ca0.3MnO3 material improvements and device measurements. IEEE Journal of the Electron Devices Society. 2018;**6**:146

[26] Tsymbal EY, Kohlstedt H. Tunneling across a ferroelectric. Applied Physics. 2006;**313**:181

[27] Garcia V, Bibes M. Ferroelectric tunnel junctions for information storage and processing. Nature Communications. 2014;**5**:4289

[28] Jeong DS, Hwang CS. Nonvolatile memory materials for neuromorphic intelligent machines. Advanced Materials. 2018;**30**:1704729

[29] Wen Z, Wu D. Ferroelectric tunnel junctions: Modulations on the potential barrier. Advanced Materials. 2020;**32**:1904123

[30] Zhang Q, Solanki A, Parida K, Giovanni D, Li M, Jansen TLC, et al. Tunable ferroelectricity in ruddlesden–popper halide perovskites. ACS Applied Material Interfaces. 2019;**11**:13523

[31] Ovshinsky SR. Reversible electrical switching phenomena in disordered

structures. Physical Review Letters. 1968;**21**:1450

[32] Yamada N, Ohno E, Nishiuchi K, Akahira N, Takao M. Rapid-phase transitions of GeTe-Sb2Te3 pseudobinary amorphous thin films for an optical disk memory. Journal of Applied Physics. 1991;**69**:2849

[33] Tominaga J, Kikukawa T, Takahashi M, Phillips RT. Structure of the optical phase change memory alloy, Ag–V–In–Sb–Te, determined by optical spectroscopy and electron diffraction. Journal of Applied Physics. 1997;**82**:3214

[34] Cassinerio M, Ciocchini N, Ielmini D. Logic computation in phase change materials by threshold and memory switching. Advanced Materials. 2013;**25**:5975

[35] Burr GW, Breitwisch MJ, Franceschini M, Garetto D, Gopalakrishnan K, Jackson B, et al. Phase change memory technology. Journal of Vacuum Science & Technology B. 2010;**28**:223

[36] Siegrist T, Jost P, Volker H, Woda M, Merkelbach P, Schlockermann C, et al. Disorder-induced localization in crystalline phase-change materials. Nature Materials. 2011;**10**:202

[37] Shariffar A, Salman H, Siddique TA, Gebril W, Manasreh MO. Resistive switching in FTO/CuO–Cu2O/Au memory devices. Micro & Nano Letters. 2020;**15**:853

[38] Cui H-P, Li J-C, Yuan H-L. Bending effect on the resistive switching behavior of a NiO/TiO2 p–n heterojunction. RSC Advances. 2018;**8**:19861

[39] Gergel-Hackett N, Hamadani B, Dunlap B, Suehle J, Richter C, Hacker C, et al. A flexible solution-processed memristor. IEEE Electron Device Letters. 2009;**30**:706

[40] Wang M, Cai S, Pan C, Wang C, Lian X, Zhuo Y, et al. Robust memristors based on layered two-dimensional materials. Nature Electronics. 2018;**1**:130

[41] Li Z, Qiao H, Guo Z, Ren X, Huang Z, Qi X, et al. High-Performance Photo-Electrochemical Photodetector Based on Liquid-Exfoliated Few-Layered InSe Nanosheets with Enhanced Stability. Advanced Functional Materials. 2018;**28**:1705237

[42] Wang Y, Wu F, Liu X, Lin J, Chen J-Y, Wu W-W, et al. High on/off ratio black phosphorus based memristor with ultra-thin phosphorus oxide layer. Applied Physics Letters. 2019;**115**:193503

[43] Patel M, Hemanth NR, Gosai J, Mohili R, Solanki A, Roy M, et al. MXenes: promising 2D memristor materials for neuromorphic computing components. Trends in Chemistry. 2022;**4**:835

[44] Xing C, Huang W, Xie Z, Zhao J, Ma D, Fan T, et al. Ultrasmall bismuth quantum dots: Facile liquid-phase exfoliation, characterization, and application in high-performance UV–Vis photodetector. ACS Photonics. 2018;**5**:621

[45] Shi Z, Cao R, Khan K, Tareen AK, Liu X, Liang W, et al. Two-dimensional tellurium: Progress, challenges, and prospects. Nano-Micro Letters. 2020;**12**:99

[46] Huang C-H, Huang J-S, Lin S-M, Chang W-Y, He J-H, Chueh Y-L. ZnO1–x nanorod arrays/ZnO thin film bilayer structure: From homojunction diode and high-performance memristor to complementary 1D1R application. ACS Nano. 2012;**6**:8407

[47] Yu Y, Wang C, Jiang C, Abrahams I, Du Z, Zhang Q, et al. Resistive switching behavior in memristors with TiO2 nanorod arrays of different dimensions. Applied Surface Science. 2019;**485**:222

[48] Kwon JU, Song YG, Kim JE, Chun SY, Kim GH, Noh G, et al. Surface-dominated HfO2 nanorod-based memristor exhibiting highly linear and symmetrical conductance modulation for high-precision neuromorphic computing. ACS Applied Material Interfaces. 2022;**14**:44550

[49] Choi BJ, Yang JJ, Zhang MX, Norris KJ, Ohlberg DAA, Kobayashi NP, et al. Nitride memristors. Applied Physics A: Materials Science & Processing. 2012;**109**:1

[50] Khalid Rahmani M, Ali Khan S, Farooq Khan M, Hee Kang M. Fully solution-processed organic RRAM device with highly stable butterfly-shaped hysteresis. Materials Science and Engineering: B. 2022;**282**:115784

[51] Nguyen VC, Lee PS. Coexistence of write once read many memory and memristor in blend of poly(3,4-ethylenedioxythiophene): Polystyrene sulfonate and polyvinyl alcohol. Scientific Reports. 2016;**6**:38816

[52] Park JH, Khan SA, Rahmani MK, Cho J, Kang MH. Enhancement of resistive switching behavior of organic resistive random access memory devices through UV-Ozone treatment. Materials Research Express. 2022;**9**:085903

[53] Wu H-C, Yu A-D, Lee W-Y, Liu C-L, Chen W-C. A poly(fluorene-thiophene) donor with a tethered phenanthro[9,10-d]imidazole acceptor for flexible nonvolatile flash resistive memory devices. Chemical Communications. 2012;**48**:9135

[54] Basori R, Kumar M, Raychaudhuri AK. Sustained resistive switching in a single Cu:7,7,8,8-tetracyanoquinodimethane nanowire: A promising material for resistive random access memory. Scientific Reports. 2016;**6**:26764

[55] Li L, Sun Y-M, Wen D-Z. The storage cell circuit with memristor characteristics of poly(N-vinylcarbazole) films. Chinese Physics Letters. 2015;**32**:078501

[56] Heo JH, Shin DH, Moon SH, Lee MH, Kim DH, Oh SH, et al. Memory effect behavior with respect to the crystal grain size in the organic-inorganic hybrid perovskite nonvolatile resistive random access memory. Scientific Reports. 2017;**7**:16586

[57] Wu Y, Wei Y, Huang Y, Cao F, Yu D, Li X, et al. Capping CsPbBr3 with ZnO to improve performance and stability of perovskite memristors. Nano Research. 2017;**10**:1584

[58] Solanki A, Guerrero A, Zhang Q, Bisquert J, Sum TC. Interfacial mechanism for efficient resistive switching in ruddlesden–popper perovskites for non-volatile memories. Journal of Physical Chemistry Letters. 2020;**11**:463

[59] Song X, Yin H, Chang Q, Qian Y, Lyu C, Min H, et al. One-dimensional (perovskite for ultralow power consumption resistive memory). Research. 2021;**2021**:9760729

[60] Park Y, Kim SH, Lee D, Lee J-S. Designing zero-dimensional dimer-type all-inorganic perovskites for ultra-fast switching memory. Nature Communications. 2021;**12**:3527

[61] Mu B, Hsu H-H, Kuo C-C, Han S-T, Zhou Y. Organic small molecule-based RRAM for data storage

and neuromorphic computing. Journal of Materials Chemistry C. 2020;**8**:12714

[62] Solanki A, Wu B, Salim T, Lam YM, Sum TC. Correlation between blend morphology and recombination dynamics in additive-added P3HT:PCBM solar cells. Physical Chemistry Chemical Physics. 2015;**17**:26111

[63] Hung C-C, Chiu Y-C, Wu H-C, Lu C, Bouilhac C, Otsuka I, et al. Conception of stretchable resistive memory devices based on nanostructure-controlled carbohydrate-block-polyisoprene block copolymers. Advanced Functional Materials. 2017;**27**:1606161

[64] Chiolerio A, Bocchini S, Crepaldi M, Bejtka K, Pirri CF. Bridging electrochemical and electron devices: fast resistive switching based on polyaniline from one pot synthesis using FeCl3 as oxidant and co-doping agent. Synthetic Metals. 2017;**229**:72

[65] Weber D. CH3NH3PbX3, ein Pb(II)-system mit kubischer perowskitstruktur / CH3NH3PbX3, a Pb(II)-system with cubic perovskite structure. Zeitschrift für Naturforschung B. 1978;**33**:1443

[66] Sum TC, Mathews N. Advancements in perovskite solar cells: Photophysics behind the photovoltaics. Energy & Environmental Science. 2014;**7**:2518

[67] Berry J, Buonassisi T, Egger DA, Hodes G, Kronik L, Loo Y-L, et al. Hybrid organic–inorganic perovskites (HOIPs): Opportunities and challenges. Advanced Materials. 2015;**27**:5102

[68] Stoumpos CC, Cao DH, Clark DJ, Young J, Rondinelli JM, Jang JI, et al. Ruddlesden–popper hybrid lead iodide perovskite 2D homologous semiconductors. Chemistry of Materials. 2016;**28**:2852

[69] Mao L, Ke W, Pedesseau L, Wu Y, Katan C, Even J, et al. Hybrid Dion–Jacobson 2D lead iodide perovskites. Journal of the American Chemical Society. 2018;**140**:3775

[70] Zhu X, Lee J, Lu WD. Iodine vacancy redistribution in organic–inorganic halide perovskite films and resistive switching effects. Advanced Materials. 2017;**29**:1700527

[71] Zhu X, Lu WD. Optogenetics-Inspired tunable synaptic functions in memristors. ACS Nano. 2018;**12**:1242

[72] Ham S, Choi S, Cho H, Na S-I, Wang G. Photonic organolead halide perovskite artificial synapse capable of accelerated learning at low power inspired by dopamine-facilitated synaptic activity. Advanced Functional Materials. 2019;**29**:1806646

[73] Zhou F, Liu Y, Shen X, Wang M, Yuan F, Chai Y. Low-voltage, optoelectronic CH3NH3PbI3-xClx memory with integrated sensing and logic operations. Advanced Functional Materials. 2018;**28**:1800080

[74] Kim D, Kim I-J, Lee J-S. Memory devices for flexible and neuromorphic device applications. Advanced Intelligent Systems. 2021;**3**:2000206

[75] Zidan M, Jeong Y, Lee J, Chen B, Huang S, Kushner M, et al. A general memristor-based partial differential equation solver. Nature Electronics. 2018;**1**:411

[76] Prezioso M, Merrikh-Bayat F, Hoskins BD, Adam GC, Likharev KK, Strukov DB. Training and operation of an integrated neuromorphic network based on metal-oxide memristors. Nature. 2015;**521**:61

[77] Alibart F, Zamanidoost E, Strukov DB. Pattern classification by memristive crossbar circuits using ex situ and in situ training. Nature Communications. 2013;**4**:2072

[78] Yao P, Wu H, Gao B, Tang J, Zhang Q, Zhang W, et al. Fully hardware-implemented memristor convolutional neural network. Nature. 2020;**577**:641

[79] Wu Y, Wang X, Lu WD. Dynamic resistive switching devices for neuromorphic computing. Semiconductor Science and Technology. 2022;**37**:024003

[80] Haupt C, Huber AB. How axons see their way--axonal guidance in the visual system. Frontiers in Bioscience. 2008;**13**:3136

[81] Saal HP, Bensmaia SJ. Biomimetic approaches to bionic touch through a peripheral nerve interface. Neuropsychologia. 2015;**79**:344

[82] Baron R, Maier C, Attal N, Binder A, Bouhassira D, Cruccu G, et al. Peripheral neuropathic pain: a mechanism-related organizing principle based on sensory profiles. Pain. 2017:158

Chapter 10

Technology Strategy Formulation for Global Corporations

Anil Vaidya

Abstract

Most global companies operate in multiple countries and locations all around the world. The focus on technology and economics is not enough as the global corporations have affiliates and subsidiaries that carry social and political undertones stemming from cultural and geographical diversity. The execution of digitalization/ technology strategy is governed by the interplay of social structures and technology. Orlikowski, Sabatier and other researchers found that social aspects significantly contribute to strategy execution success. Lindbolm, Dunshire and others have brought into limelight the bipartisan politics, mutual adjustment etc. This chapter discusses major elements of technology strategy viz. Technological considerations, Operational considerations, Economic aspects, Social aspects and Political issues. Converging, the chapter presents aspects of technology strategy of national companies and multinational companies, specifically focusing on social and political angles of a multinational companies. Third part focuses on technology strategy formulation process, concluding that it is the strength of the process of strategy formulation that will help the companies in the changing dynamic environment business scenario. The model offered here is titled IPCRC model adapted from Vaidya that addresses meeting socio-political challenges of a multinational company. Conceptually this model may be used in any business or functional strategy formulation.

Keywords: technology strategy, global corporation, multinational company, technology strategy formulation, IPCRC model

1. Introduction

Businesses have been making changes, transforming the way they work for several decades. Globalization set in bringing changes in the way businesses work. The labor, capital and goods started moving around the globe with much ease, along-with the work itself. The service industry came to fore. The pace of change has only been accelerating, much more so in the past decade. At the nucleus of these changes have been the technological progress. The internet, proliferation of mobile, cloud, social networks all had a huge impact on the relation of customers with businesses. New media became available for interaction. Wynn and Jones [1] elaborated the IT strategy in the digital world in a hotel industry. The researchers here focus on changed customer requirements and resulting functionality that needs to be incorporated. Mankins et al. [2] advocates that emerging from pandemic in the digital world the companies should get their employees to work with technologies in a coordinated way. The Covid-19 pandemic forced businesses and customers to adopt new ways of technology advances and internalization. The technology's position in business and society became more visible. In the earlier years the technology was always being treated as a supporting function. Today almost no business or society can avoid use of technological implants, they have taken the center stage. Gerard et al. [3] have discussed the changes in technology and innovation on account of pandemic and how it has changed the society and businesses.

Technology strategy has been a topic of discussion in the practice. Many researchers and practitioners have written about 'alignment with business' and 'adoption of technology'. While these concepts were right at the time of writing, over past few years there are considerable changes in the business practices. Past 2 years of pandemic demonstrated accelerated use of technology. Pereira et al. [4] study documents the business transformation brought by the internalization of digital technologies before and during pandemic. From that perspective businesses do not discuss alignment or adoption, accepting that the technology has been integral to running of businesses. Technology strategy is not considered as a support structure, instead it is now embedded in the business strategy. For instance, in the world of Finance and Logistics it has taken center stage occupying seats at the strategy councils of corporates. I will discuss here factors that influence Technology strategy. I have divided the discussion in three major parts viz. Technology Strategy of National companies, Technology Strategy of Multinational Companies and Strategy Formulation Process.

2. Part I: technology strategy of national companies

To make it clear I bring here concept of a National Company. It is the one that has all businesses concentrated within the national boundaries, geographical and judicial. Within these boundaries company has employees and customers who share similar culture, language, traditions etc. In the political world such countries are termed as Nation-States, in the contemporary world most countries fall in this category.

Here I present an overview of the elements that are addressed by the technology strategist.

1. Technical considerations

2. Operational considerations

3. Economic aspects

One needs to remember that although one can discuss these as individual elements they are highly interdependent. A change in one will have ripple effect on the second and will need to change and so on. Yeh et al. [5] studied importance of information system capability as a factor in the success of e-business information technology strategy. The capability refers to both technical and operational aspects of technology. The researchers concluded that such a capability has direct impact on the e-business. Saghaeiannejad-Isfahany et al. [6] researched feasibility of telemedicine in the Isfahan province. They found that medical professionals were in favor of such a system and considered it useful, expressing view that the technical and operational feasibility was required to make investments.

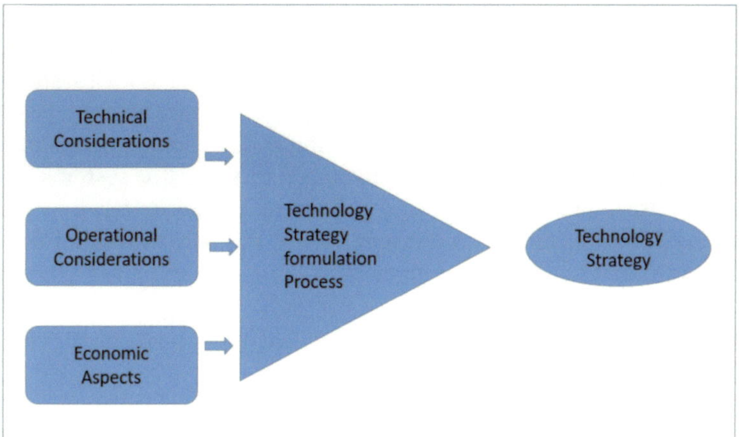

Figure 1.
Technology strategy of a national company.

Figure 1 shows three elements that are considered while formulating technology strategy. It depicts the technical considerations, operational considerations and economic aspects that need to be taken into account for technology strategy. The technology strategy itself is an outcome of strategizing process. For the sake of understanding, we discuss all three here.

2.1 Technical considerations

While technology developed at a rapid pace many solutions and applications became available to businesses. The ERP, CRM, MRP, Warehouse management etc. working on-premises have been implemented for a long time, the internal data centers and the support teams became part of organization structure. Then the cloud came into picture, enticing companies move to clod based applications. This was an opportunity for the people to rethink technology strategy moving entire or part infrastructure to cloud. Simultaneously internet based success based stories started emerging such as Amazon, Netflix and Spotify. The e-commerce offered a different route to customers.

Today one needs to account for multiple facets of technology:

1. On-premises vs. cloud based systems

2. Support team organization

3. Specific business requirements such as reach to customers, suppliers and employees

4. Business criticality of open vs. closed systems e.g. should some parts of system be opened to vendors and customers

5. Security aspects thinking about research & development and innovation departments digitization

6. Geographical reach of business such as operating in multiple states/regions/provinces of a country and the compliance requirements of each.

Points 3–6 above are important for the technology strategy, they relate to business direction and the requirements. As an instance, some forward looking businesses may want to have their customers and vendors on their own internal ERP systems. Some others may consider internal system to be strictly opened only to employees of the organization. Another case where the engineering and construction business may want to work with their contractors and subcontractors on same systems for design and project monitoring system, while keeping financial data strictly kept confidential to the company. Besides, considerations may be substantially different for a national organization from that of a local organization.

2.2 Operational considerations

The operational considerations are related to technological considerations. While having an in-house data center will need the support structure such as facilities, people and space. The support considerations involve employing own people. That would mean getting correct talent in the organization and retaining it. In the technology space generally opportunities galore resulting in higher turnover of the people. It becomes a major human resource exercise to manage such talent. The technology advances also affect the skills and competency requirement.

On the other side one may consider outsourcing some of these services. If the cloud based managed services are engaged then internal technology human resource requirement go down. Further the managed service vendor guarantees the uptime and quality of service under 'Service Level Agreement (SLA)'. In such situations the technology department/division may require people that are experts in outsourcing monitoring and control.

Consolidating all technology resources in one central locations helps to lower costs, however operationally servicing customers, employees and vendors in other locations to their satisfaction may be challenging. A country such as China, USA and Russia are examples where they operate in different time clocks across regions. That results into providing service from central system to match necessary clocks. It is of course possible to do this and there are companies that operate such centralized systems. Besides, the large companies operate central services such as Human Resources, Finance and Technology. Their locations add the complexity of providing service from central technology system. Mahamed et al. [7] describe a technology based solution titled 'home-based maternal record (HBMR)' explored in Zimbabwe. The researchers have assessed operational feasibility of providing a simple solution in rural areas to improve health record.

There are always possibilities of acquisitions, mergers and divestures. When the businesses merge or make acquisitions there is certain planning that can be done in advance in the technology strategy to ensure smoother integration of businesses. In case of divesture of some business lines or sectors the technology strategy changes turn out to be much more challenging. This is more so as the divesture is announced only when all legal procedures are completed and the necessary approvals are acquired. All these require responding technology matching the timeline set by these events.

2.3 Economic aspects

A central expectation of the technology implementation is that it creates value for business. The 'value creation' is absolutely vital for investing in technology.

The typical financial parameters such as Return on Inventor (ROI), Net Present Value (NPV), Discounted Cash flow and Payback period are the metrics used for evaluating projects in most of the organizations. Strassmann [8] advocates the use of ROI to make decisions on IT investments adding that economic criteria are important. The decision makers should way the best and worst case scenarios. The investment in technology projects is approved only if it means certain level of returns. Many businesses today would like to be asset light, that mean they may not want data center on their balance sheet. Such companies may prefer to host their software on the cloud or another data center. Even in such as meticulous calculations are done to understand impact of moving capital expenditure (capex) operational expenditure (opex) in the long run and short run. Today many other inputs need to be taken into account well beyond financial metrics, some such examples, amongst others, may be compliance requirements, nice to have technology and threat mitigation.

Smith et al. [9] bring out importance of business value creation through technology strategy, further specifying that the business and technology strategy need to complement each other to generate value. **Figure 2** picturizes examples of value of technology as may be considered by different people in different context. The decision makers may have different considerations of value generated by technology. Efficiency and time saving are the most common values. One also needs to distinguish between 'nice to have' and 'essential to have'. For instance, $200 smart phones for the sales people traveling in the field may fall in the category of 'essential to have' however, smartphones of $600 may be termed 'nice to have'.

One may think of WhatsApp acquisition by Facebook in 2014 as the one beyond pure financial considerations, as more and more people started using WhatsApp. Skype acquisition by Microsoft in 2011 was not making much of money then. EMC by Dell and 2016, Autonomy by HP in 2011 all showcase high profile acquisitions. While one

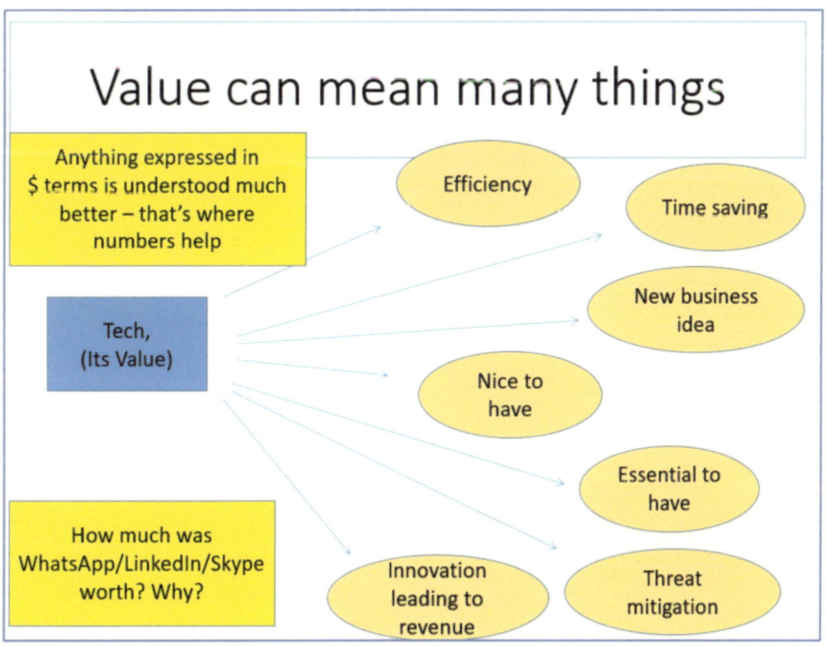

Figure 2.
Value of technology.

may argue that these were decisions at the enterprise level having significant business considerations, one cannot dispute the fact that all acquired companies had special technology that was valued by acquiring company. WhatsApp and Skype were clear examples of new ways a communicating, to the extent that Mark Addreessen in his famous paper "Software is Eating the World" in 2011 classified Skype as telecom company. Silvius [10] discusses value of technology, also questions mindless applicability of ROI in every technology investment. In a way he brings out salient, even intangible, outcomes of technology. Brynjolfsson and Hitt [11] found in their study that benefits of information technology investments are difficult to measure and are often intangible. However, there is strong link between information technology and productivity gains.

It is important to also consider the technology portfolio management of company where acceptance/rejection/postponement of technology projects is done. Besides other metrics the financial metric of vale generated by technology carries high weight. Let us consider here simple examples of Finished Goods Inventory management and Accounts Receivables by a fast moving consumer products company. The computation shown here is oversimplified to give first level of understanding.

Scenario 1: An FMCG company operating within a national boundary has revenue of $10 million, with COG (cost of Goods) at 60%. It generally maintains 30 days of finished goods (FG) inventory. An IT company offers to provide an IT solution that can reduce FG inventory by 15% at a cost of $20 thousand per annum. Given inventory carrying cost of 20% should the solution be accepted?

Table 1 shows computation of savings possible by reduction in finished goods inventory.

One needs to compute savings in cost on account of reduction in FG inventory. Carrying finished goods inventory entails costs such as bank interest, insurance, space, administration and obsolescence. The scenario given here indicates such inventory carrying cost to be 20%. The level of such cost varies based on the kind of products, the interest changes, location of warehouse etc. As an example obsolescence may be very high in perishable food items but low in metals.

1	30-days FG inventory at revenue	= $10 m/12	≈ $833 thousand
2	30-days FG inventory at COG	= $833 K × 0.60	≈$500 thousand
3	Reduction of FG at 15%	= $500 K × 0.15	= $75 thousand
4	Inventory Carrying cost reduction in FG at 20%	= $75 K × 0.20	= $15 thousand
5	Cost of IT solution		= $20 thousand

Table 1.
Scenario 1: Computation of savings.

1	45-days accounts receivables	= $(120 m/12) × 1.5	≈ $15 million
2	30-days accounts receivables	= $ 100 m/12	≈$ 12 million
3	Reduction of AR at 10%	= $ 15–12 m	= $3 million
4	Reducing cost of financing at 10%	= $3 m × 0.10	= $300 thousand

Table 2.
Scenario 2: Computation of savings.

In the above computation adopting the solution at $20 K gives no benefit to the company. Solution will not be accepted.

Scenario 2: Consider a company with a turnover of a $120 million selling products in a single country, with a 45 day accounts receivables. An IT solutions provider offers a service that would bring down the accounts receivables to 30 days. Given the cost of financing the receivables at 10% what is acceptable cost per year of IT service?

Table 2 shows computation of savings in this scenario.

Accounts receivables show outstanding amounts to be received from customers. Generally, the selling company extends credit to the customers. In this scenario the company's AR is equivalent to 45-days sales. As the receivables are financed by bank loan it is in the interest of the selling company to minimize it as much as possible. Bringing down AR from 45 to 30 days saves $300 thousand for company, as seen in **Table 2**. In this case company will accept an IT service if it costs less than $300 thousand per year.

3. Part 2: technology strategy of multinational companies

The globalization set pace for expansion of those who could sell their products and services around the globe. Netflix, Amazon, Google, Mondelez, Vodafone, McDonald and may such companies in diverse sectors established their businesses in multiple countries. Challenges are different now which result from different cultures, traditions, languages from customer service perspective and the compliance requirement from local regulations perspective. Geppert and Mayers [12] bring out the socio-political dimension of such organizations. All these became part and parcel of the technology strategy as well.

Two other very relevant factors, that technology strategists unaware of, came to fore viz. Social and Political environment. Adding these two:

1. Technical considerations
2. Operational considerations
3. Economic considerations
4. Social factors
5. Political environment

Figure 3 shows the factors that need to be considered while formulating technology strategy of multinational company. Do notice, in addition to technical, operational and economic aspects, socio-political dimension has been added as an intervening factor. I will discuss here additional two factors added above and then move onto the changes in other factors.

Operational considerations are influenced by social and political factors in a multinational company. They are crucial for a company having manufacturing and/or service centers in multiple countries and customers spread globally. The technology strategy of such a company is fundamentally different for a multinational organization from that of a national organization. I have discussed operational considerations further in Section 3.3.

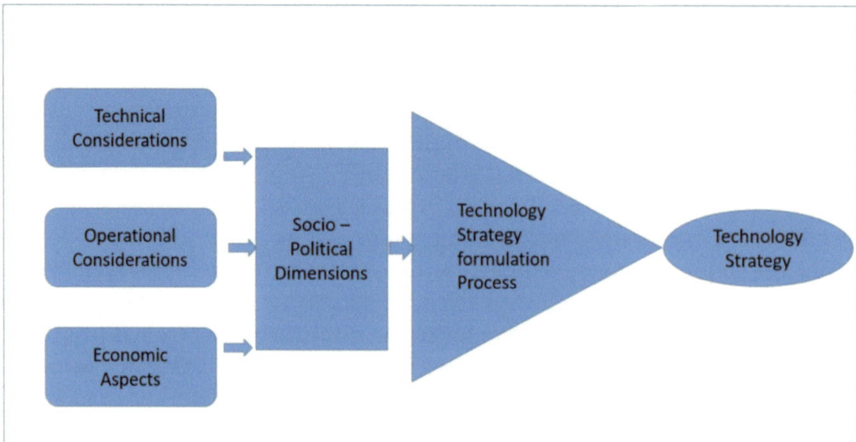

Figure 3.
Technology strategy of a multinational company.

3.1 Social factors

Societies in different countries have different level of adoption to technology. As an example World Bank published Digital Adoption Index for 180 countries based on data available up to 2016. The Statista gives Digital Competitiveness Ranking for 2021 representing countries ability to adopt digital technologies. Having such information handy is helpful in deciding depth of technology that may be planned in certain countries.

Orlikowski [13] concluded many decades ago that the use of technology is ultimately depended on the users of technology and not on the designers of technology. Same technology may be deployed in completely different manner by one set of people than others. An excellent example may the use of intra-company chat systems. There are companies who have used it as a tool for solving complex technical issues with the collaboration of experts spread around the world, while others have used it only for social chat. Orlikwoski and Gash [14] talk about 'Technological Frames' that users develop. Their understanding and use of technology is influenced by their technological frames.

In the contemporary world the technology strategy of multinational company is invariably formulated in the parent company, as most other business/functional strategies. It is expected that the affiliate companies and business unit spread around the world accept and follow the strategy as mandate. The conflict originates in the expectation that people working in different cultures in the affiliate company would simply accept what is conveyed to them. In the real world many times the perceived impact creates dissatisfaction, also resulting in low motivation and lower morale. The head of such a far-away unit understands and raises it at a higher level of the parent management, which now turns out to be a political issue as well shown in **Figure 4**.

Figure 4 shows the how a social factor can turn to be a political issue. It may be just the perception of people amongst the employees of affiliate company that may have to be overcome at political level. Orlikowski [15] brings out interpretive flexibility of technology. This allows people in affiliates to interpret social implications which the interest groups can shape into political issue quickly.

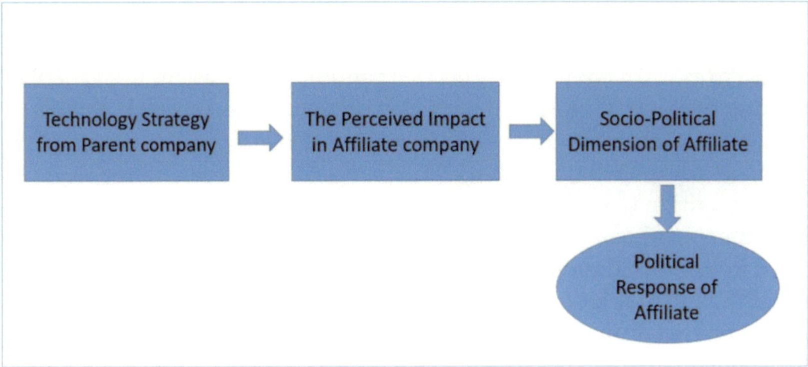

Figure 4.
Socio-political environment of a multinational company.

3.2 Political environment

Politically there are many facets that one has to consider while providing technology services to the stakeholders in that country. Some example here:

1. Many countries now specify that the data of residents of the country be stored on servers that are located within the boundaries of country.

2. United Nations publish list of fragile states that potentially do not have stable social, political and/or economic conditions.

3. FATF (Financial Analyst Task Force) a body set up by FinCEN (Financial Crimes Enforcement Network) of US Treasury publishes a list of countries that do not satisfactorily meet restrictive measures that prevent financing Terrorism and Money laundering.

4. There are many instances of the inter and intra state conflicts that have been seen largely in Asia and in South African countries. These countries also account for large pie of total population and the market potential.

Besides such external factors there are internal factors that influence technology strategy. A multinational organization operating in five continents generally has the manufacturing plants in low cost countries, the Research and Development in high talent rich locations and the back office work in most optimum cities. The resultant countries have a sort of unwritten hierarchy in the multinational organization. The heads of those countries have ears to the higher ups in the company and they have a say about every possible strategic and tactical plan of the company. Strategy formulation that happens in such a multi-country, multi-product, multi-division matrix organization can be best described in Lindblom's [16] terminology as 'Partisan Mutual Adjustment'. The strategist needs to manage certain give and take while making final plan. The various groups in a multinational company working in affiliates and parent company have own interests. The strategist needs to balance these interests, the process is referred as partisan mutual adjustment, typically found in the political circles.

One may also witness an intervention termed as collibration by Dunshire [17]. Collibration is the term used by Dunshire to indicate a higher level intervention such as a directive to award a contract to a certain vendor, or to appoint to particular person in a particular position. Philips deal with Dell on outsourcing of global desktop services that was canceled in 12 months in 2006, later on awarded to IBM. Dell was also supplier of components to Philips. One may wonder if the desktop deal with Dell was part of condition to maintain customer-supplier relationship.

Thirdly there has been a concept of coalition that is usually seen in political circles. A strategist in a multinational company needs to aware and work with coalition actors as described by Sabatier [18]. The coalition politics, also referred s 'Realpolitik' is a term that represents groups of people joining hands in a given context without any regard to their own ideology or own goals. It is also called as pragmatic politics. In a multinational company one may find such coalitions that exit only for a transient period to serve certain common goals.

3.3 Multinational company: operational considerations

1. A multinational company operating in multiple countries has people working in various locations. Invariably people work in different time zones and follow different clocks. The strategist needs to be conscious of the impact of decision about centralizing and decentralizing systems. The centralized system gives an advantage of uniformity and cost reduction, but also increases challenges of satisfying requirements of diverse set of people with diverse requirements.

2. Countries like Japan and China have further requirement of local language software. Certain amount of relief may be providing by using standard systems such as SAP and Oracle. However, they will still to be integrated with software for local needs.

3. One may do better by having a proper mix of centralized and decentralized systems.

4. It may be economical to have India, Pakistan and other subcontinent countries to work on same system. These countries have similar cultures and tradition, however locating the central system either in India or Pakistan may pose a problem as the political environment is never conducive to such a practice. As a result, the central systems needs to be in a third country increasing the costs.

4. Part III: strategizing technology is a process, not an event

It is quite a challenging tasks to develop technology strategy of a large multinational corporation. In recent years the changes on all fronts, the economic, political, social and environment, are extremely rapid. These changes impact businesses and their long-term and short-term strategies. Just most recent global catastrophic event, the pandemic, has affected and impacted entire population of the globe and consequently businesses around the globe. The painful 'Brexit' has created newer issues in Europe and countries dealing with UK. Add to this the Ukraine invasion of Russia. The resulting inflation and shortages of various necessities including energy has posed completely new perspective of the world to business leaders.

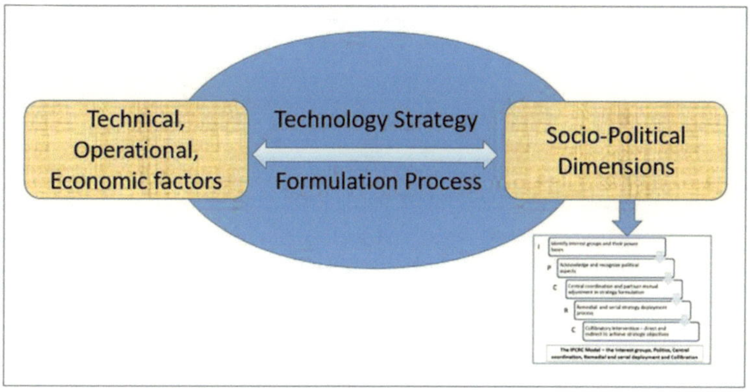

Figure 5.
Two important input blocks to strategy formulation process.

The Technology Strategist cannot isolate herself from this new rapidly changing world full of unanticipated events. The strategy formulated by her has to be:

1. Flexible, nimble, easy to be reformulated

2. Easy to execute

3. Minimum possible lock-ins

4. Balancing needs of organization and those of technology resources

It is critical to attend to the process of strategy formulation, rather than just the strategy itself. I propose to focus on the strategy formulation process instead of final product as technology strategy. Letza et al. [19] emphasize that no business process can be isolated from power, social relations and organizational context. I present here the Technology strategy formulation process in two blocks. First block to complete initial assessment of Technology and Economic aspects, followed by the second block adapted from Vaidya [20].

The two blocks shown in **Figure 5** indicate important inputs that are required for Technology Formulation Process of a multinational company. The left block of Technical, Operational and Economic considerations was discussed in part I, here we will explore the second block of Socio-Political dimensions. In order to get clarity, I have exploded the second block in **Figure 6**.

IPCRC model in **Figure 6** is an interesting model that takes into account social and political interests of a multinational company. It caters to the socio-political environment of a multinational company, proposing the activities that need to be completed while developing technology strategy. Here is a short description of each activity.

I—Identify the interest groups and their power bases: the interest groups carry their own norms and ways of work. It is important for strategist to understand the power bases behind such groups. As an example the largest business division in a multinational company may have higher power to influence decisions.

P—Acknowledge and recognize political aspects: the affiliate company heads and senior managers may have own ambitions. It is useful to know these persons and their thoughts.

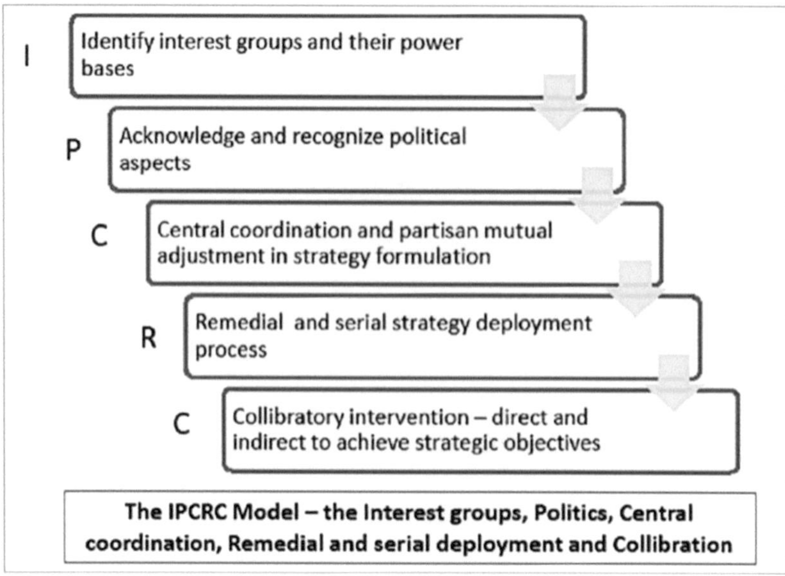

Figure 6.
IPCRC model adapted from Vaidya [20].

C—Central coordination and mutual adjustment: it is useful to know the coalitions and their requirements beforehand. The give and take with necessary adjustments rather than mandate produces better result.

R—Remedial and serial strategy deployment: a step by step implementation proves better than the big bang approach. It allows any remedial measure to be incorporated as one progresses.

C—Collibratory intervention: it is always a possible that the higher level intervention has to be accommodated, such as a directive to place a data center in a particular region or having a preferred network provider in a country. The strategist needs to make room for such a possibility.

By following this IPCRC model a strategist will be able to develop an agreeable strategy which becomes easier to execute in various cultures and geography. Considering deployment challenges at an early stage of strategy formulation results in a executable strategy.

Since 2020 the world has faced pandemic, shortages of supplies, human tragedy, the oil shock, the war between Russia and Ukraine, the emerging tensions in Asia, many such worrisome issues. As a result, the world order, the relations amongst various sovereign states are being redrawn all the time. The globalization seems to be giving way to nationalization, there are people who believe that deglobalization is already in offing. In the dynamic and ever changing geopolitical and national and social interests it is extremely critical that the strategist focuses on strategy formulation process. The product, the strategy, is important however one has to build ability to quickly reform it using right kind of processes.

5. Part IV: conclusion

It is important to realize that the contemporary world faces rapidly changing, fast pace dynamic scenario on all fronts. Technology Strategy has to be nimble and flexible to match the dynamism of businesses, which by itself depends on the ever transforming environment. In the past the national companies could rely on three known aspects of technical, operational and economic considerations. However, the spread of global companies around the world in multiple countries brought in focus the socio-political aspects of such a company. Add to it the ever changing dynamic geopolitical situation riddled with supply issues, state conflicts, inflation and nationalization policies. The way forward here is to strengthen the strategy formulation process, rather than focusing on strategy alone. It needs to be seen as a continuous process instead of an event or an output. One needs to follow the robust technology formulation process suggested here.

Contribution to body of knowledge

This chapter identifies the challenges faced in a multinational company. The characteristics of a global company spread in multiple countries and cultures are very different from that of a national company. One needs a practical and pragmatic approach to ensure that technology strategy so developed is executed on the ground. Ashurst et al. [21] discusses practice orientation that relates to informal organization accentuated by knowledge and behaviors of individuals. Anticipating responses is key to a good strategy, ultimately a technology strategy needs to be agreeable not necessarily ideal in today's multilateral world order. Whittington [22] advocates shift to practice perspective as key, elaborating that the manager has to develop competence of strategist. Similarly, Sminia [23] discusses the process of layered discussion. It is understanding of people interests and impact of technology strategy on lives of people around the globe that brings the strategy deployment possible. Therefore, key is to orient oneself to practice rather than ideal in the early stages of strategy development. This chapter highlights the practice of being attentive to the socio-political factors in multinational company.

Author details

Anil Vaidya
S.P. Jain Institute of Management and Research, Mumbai, India

*Address all correspondence to: anvvaidya@gmail.com; anil.vaidya@spjimr.org

References

[1] Wynn M, Jones P. IT strategy in the hotel industry in the digital era. Sustainability. 2022;**14**:1-14

[2] Mankins M, Garton E, Schwartz D. Future-proofing your organization. Harvard Business Review. Sep-Oct 2021

[3] Gerard G, Lakhani K, Puranam P. What has changed? The impact of COVID pandemic on the technology and innovation management research agenda. Journal of Management Studies. 2020. DOI: 10.1111/joms.12634

[4] Pereira C, Veloso B, Durao N, Moreira F. The influence of technological innovations on international business strategy before and during COVID-19 pandemic. Procedia Computer Science. 2022;**196**:44-51

[5] Yeh C, Lee G, Pai J. How information system capability affects e-business information strategy implementation: An empirical study in Taiwan. Business Process Management Journal. 2012;**18**(2):197-218

[6] Saghaeiannejad-Isfahany S, Jahanbakhsh M, Shaya A. Telemedicine operational feasibility in selected hospitals of Isfahan University of Medical Sciences in 1393. International Journal of Health system & Disaster Management. 2015;**3**(4):189-194

[7] Mahomed K, Mason E, Warndorf T. Home-based Mother's record: Operational feasibility, understanding and usage in a rural Community in Zimbabwe. Tropical Doctor. 2000;**30**:155-159

[8] Strassmann P. Why ROI ratios are now crucial to IT investment. Butler Group Reviews. 2002:5-7

[9] Smith H, Mckeen J, Singh S. Developing information technology strategy for business value. Journal of Information Technology Management. 2007;**XVIII**(1):49-58

[10] Sinvius A. Does ROI matter? Insights into the true business value of IT. Electronic Journal of Information Systems. 2006;**9**(2):93-104

[11] Brynjolfsson E, Hitt L. Beyond computation: Information technology, organizational transformation and business performance. Journal of Economic Perspectives. 2000;**14**(4):23-48

[12] Geppert M, Mayer M. Global, National and Local Practices in Multinational Companies. London: Palgrave Macmillan; 2006

[13] Orlikowski W. Using technology and constituting structures: A practice lens for studying technology in organizations. Organization Science. 2000;**11-4**:404-428

[14] Orlikowski W, Gash D. Technology frames: Making sense of information technology in organizations. ACM Transactions on Information Systems. 1994;**12**(2):174-207

[15] Orlikowski W. The duality of technology: Rethinking the concept of technology in organizations. Organization Science. 1992;**3**(3):398-427

[16] Lindblom C. The Intelligence of Democracy. New York: The Free Press; 1965

[17] Dunshire A. Manipulating social tensions: Collibration as an alternative mode of government

intervention. MPIFG Discussion Paper. 1993;**93**(7):2-49

[18] Sabatier P. Political science and public policy. Political Science and Politics. 1991;**24**(2):144-147

[19] Letza S, Kirkbride J, Sun X, Smallman C. Corporate governance theorizing: Limits, critics and alternatives. International Journal of Law and Management. 2008;**50**(1):17-32

[20] Vaidya A. The implications of organizational context for information systems and technology strategy formulation. In: A Study of Socio-Political Factors in Global Corporations [Thesis]. Bradford, UK: University of Bradford; 2011

[21] Ashurst C, Doherty N, Peppard J. Improving impact of IT development projects: The benefits realization capability model. European Journal of Information System. 2008;**17**:352-370

[22] Whittington R. Strategy as practice. Long Range Planning. 1996;**29**:731-735

[23] Sminia H. Strategy formation as layered discussion. Scandinavian Journal of Management. 2005;**21**(3):267-191

Chapter 11

Comparative Study of Information Security in Mobile Operating Systems: Android and Apple iOS

Fernando Cassinda Quissanga

Abstract

This theme addresses information security in mobile operating systems: Android and iOS, compare vulnerabilities, and propose security policies, the methodology is qualitative based on bibliographic reference, documentary, and quantitative tech niques, which allowed to evaluate the study in detail, the data collected in a systematic way, investigating security measures in mobile phones. It was performed the search to Google Academic® and the Portal de Periódicos Capes. As for the data analysis, it was done in a descriptive way. However, it was concluded that the iOS system presents greater information security compared to Android. Virus infections are not of concern at the moment, but other cyber attacks, spywares capture confidential information by taking to the remote server. The attacks can be identified by the victims online, in industrial, and by targeting them individually. Android and iOS operating systems are preferred by cybercriminals. As for cyber attacks, social networks are the focus. For security policies, we propose preventive, detective, and corrective measures, assign difficult passwords, use modern WPA encryption, Noroot firewall, the Privacy Badger tool for blocking websites, Panopticlick for test web, VeraCrypt to encrypt documents, and the Signal app that encrypts calls.

Keywords: information security, operating systems, mobile telephony, Android, iOS

1. Introduction

The operating systems Android and Apple iOS are the focus of cyber criminals, the comparative study of the same is of extreme importance because we can know their vulnerabilities and propose methods, policies, and security systems to protect data, mobile phones are small and embedded devices, with peculiar characteristics, allow text messaging and multimedia, because it has memory, processor, and an operating system, which allows greater use of confidential information, handling, and exchange of data between users, videos, and other collaborative work with various social networks, which enables failures in cyber security, the loss and theft of information, due to cybercrime that increases substantially, the mobile phone is vulnerable to cybercrime. The current researches focus on information security that has been

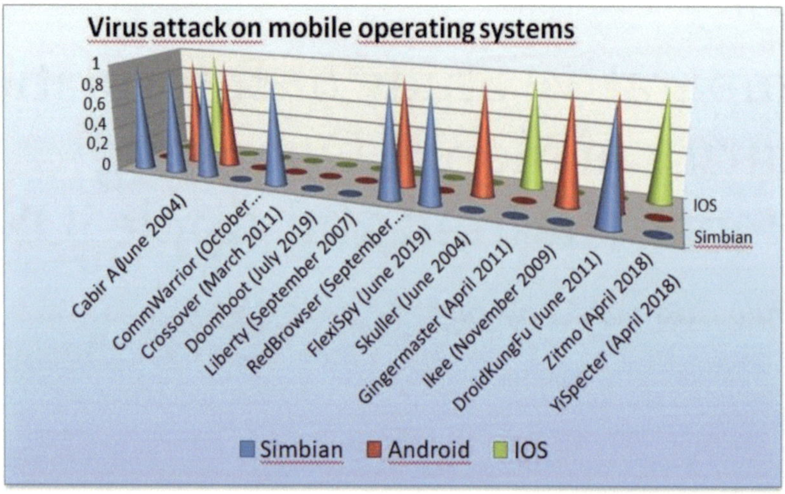

Figure 1.
Virus attack on mobile operating system. Source: Quissanga [1].

very questioned about information theft that every day increases due to the greater number of users without information security education and use mobile devices. The operating systems such as Android and iOS from Apple are now the targets of attackers, besides being very common and this feature makes them easy prey, study of this kind for our society is extremely important, we need to have basic culture of information security, control of our passwords, email, social networks, malicious software, and spies, are the apex of the modern age and information, we must be very careful to avoid irreparable losses.

However, we know that security for phones differs in relation to computers, in other words, viruses also differ in the form of attack, although there are compatible operating systems, but they differ in this perspective in which we refer, in the case of the Android operating system virus does not attack the iPhone system for a very simple issue, its core (*kernel*), that is, in the creation of systems there are different codings and languages and this way the virus is also made of programming language, to happen the attack needs to be made in the same language of the operating system, for this reason, it is important to study them separately (**Figure 1**).

The choice of this topic was in order to know which of the operating systems Android and Apple iOS is more secure and to propose security measures to protect information, since we face many problems with data protection.

2. Methodology

This study was possible through my experience as a teacher in the discipline of information security, which allowed to make a systematic evaluation of the research, so we opted for qualitative research based on bibliographic reference, documentaries, and with quantitative techniques, which allowed to evaluate the study in detail, understand how is the care of existing information, present which of the operating systems is the safest as to data protection. Detect system vulnerabilities and present some contingency measures of cyber attacks.

2.1 Data collection

The data collected systematically, investigating safety measures in mobile cellular telephony. The search was conducted in databases such as Google Academic® and Portal de Periódicos Capes. The current research is done in searches to Google Academic®, as described by Creswell [2]:

> *It is a free database that provides breadth in the search for literature from various sources, such as theses, abstracts and articles, with the advantage of being able to obtain them in full. The Capes Periodicals portal was chosen because it offers access to articles from our Portuguese repertoire selected from over 15,000 international, national and foreign journals and 126 databases with abstracts of documents in all areas of knowledge (Portal de periódicos capes).*

Researches of the type generally used the collection of data. We also used the documentary research for living moments in question and being researcher already some time on the thematic of computer security, distributed system, operating systems. It allowed to know to find some expected results. Thus, it is pertinent the quote of Chaer, Diniz, and Ribeiro [3] "as the research technique composed by an approximately high number of questions presented in writing to people, having as objective the knowledge of opinions, beliefs, feelings, interests, expectations, experienced situations, etc."

2.2 Data analysis

Data analysis was performed descriptively, allowing us to assess, compare, and organize important aspects of the research. Data analysis regarding the comparative study of operating systems, in addition to being users of these systems, we have other academic and scientific knowledge acquired during training that led us to a comparative study in specific terms, to make a deep analysis of each operating system to find which one best responds to cyber attacks. It is notable that a reduced number of works on this theme, which makes the investigation and comparison with the characteristics of the results difficult.

3. Computer viruses in cellular mobile telephony

In today's society, viruses and other malware have brought countless economic and financial losses. Their study is essential to prevent losses, the Android operating system and Apple iOS are the focus of our study, being the treatment of information thefts have been more by spyware, which sends information to a remote computer and thus be able to have all the information from the phone, banking credentials is the greatest interest.

Quissanga [4] defines:

> *[...] computer viruses in mobile phones, which we can define as a malicious software made by programming language that infects the operating system and is hosted in the program and replicates to other parts of the system, corrupts and prevents the normal operation of the software or program from starting.*

However, phone viruses are not well known, some are more destructive, others less so, but all with the aim of obtaining confidential information from users without

No.	Virus/worm name/year (updated)	Operating system
1.	Cabir A (June 2004)	Symbian
2.	Caballo de Troya (March 2017)	Symbian, Windows, Android and Mac OS X
3.	CommWarrior (October 2018)	Symbian and Android
4.	Crossover (March 2011)	Windows Mobile
5.	Doomboot (July 2019)	Symbian
6.	Liberty (September 2007)	Palm OS
7.	RedBrowser (September 2017)	J2ME
8.	FlexiSpy (June 2019)	Symbian and Android
9.	Skuller (June 2004)	Symbian
10.	Gingermaster (April 2011)	Android
11.	Ikee (November 2009)	iPhone OS (IOS)
12.	DroidKungFu (June 2011)	Android
13.	Zitmo (April 2018)	Symbian, Android
14.	YiSpecter (April 2018)	iPhone OS (iOS)

Source: Quissanga [1].

Table 1.
Computer viruses in cellular mobile telephony.

them realizing. For this reason, it is important to know the techniques used by cyber-criminals, as you cannot be too careful. There are viruses with peculiar characteristics of biological viruses, they attack and weaken the boot system preventing its normal operation. However, the attacker has many skills to be able to use these fast viruses that we will mention later in **Table 1**.

However, it is important to make mention the behavior of cyber criminals, as Futurelearn [5] tells us:

> *Talk Talk's cyber attack saw the personal details of 157,000 customers, including credit card details, leaked in October 2015. As a result, the company lost around £60 million and more than 100,000 customers, but customers were also open to potential identity fraud: in some cases, fraudsters used the data to enable them to own up as Talk Talk engineers, contacting customers and persuading them to install malware on their machines.*

In this research, we have the idea of differentiating in a very simplistic way the security between the Android and Apple iOS operating systems, which have been much discussed regarding their security, as it is already known that they are the most used systems, but we think that no matter the social status, everyone can have an Android phone or iPhone. In this way, the attacker will prepare himself for the creation of malicious "software," by knowing properly the programming language and the critical places, vulnerable to take advantage of any precise information that suits him. However, it is necessary to create security mechanisms both at the company level and at the user level. Sometimes we wonder who is responsible and we always look for a guilty party, we should think that information security is a worldwide concern, it can bring states problems, nations can fight, and companies can go bankrupt, because of information security failure. It is known that there are irreparable damages, some are moral, psychological, ethical, cultural, material, and financial. When information is leaked that is classified as legal or state secret, it can facilitate or favor one party and harm another. Previous studies do not show us conclusively that there is a real protocol to follow for the security of information in these operating systems, just as there is no best security strategy, so the most important thing is preventive measures. Because the moment any system is launched on the market is exactly when cyber criminals study the vulnerabilities for the attack.

The spread of viruses in mobile phones today is already a reality, several complaints made by users are noted, as well as mobile phone companies, concerned with creating antivirus to disseminate these virtual plagues. It is possible to verify some forms of virus transmission, which can be: by email through social networks, sometimes appearing as advertising, contact with images of doubtful origin, infected videos, visits to suspicious websites, unprotected wireless network without a secure password, the installation of malicious *software, the* use of infected memory card as secondary infection, although there is a difference for each operating system regarding the form of virus transmission due to its *kernel* (core). According to Martinelli [6], "Every operating system has a core called *kernel* which delimits its functions. It is one of the reasons that makes a mobile virus not spread easily to other devices, due to the different versions and internal structure of the various *mobile* operating systems."

Faced with this reality we see that it is not always the user who fails, but also some companies do not properly protect their customers' information.

In this respect we propose some viruses in mobile phones, as to the operating systems, their origin and attacks on the respective systems described in **Table 1** by Quissanga [1]:

> *The viruses in mobile phones originated very recently, so given their specificity, they are not so well known, but in* **Table 1** *we prefer to mention some peculiar characteristics regarding the operating system they attack. This table mentions the different viruses, their origin, the year in which they were discovered, as well as which operating system they attack, as can be seen in the first Cabir A virus, discovered in 2004 by the company F-Secure, but it attacked the Symbian operating system, after which the fastest viruses that attack Android and Apple iOS began to appear.*

However, Le Thanh [7]:

> *Although malware is growing rapidly, several ordinary users with easy access to their smartphone lack a basic understanding of the potential danger. Therefore, we need to classify samples according to similar characteristics, as well as collect more new malware to create malware families.*

4. Mobile operating systems: android and Apple iOS

4.1 Android

The Android operating system is a platform developed by the Google group designed for mobile structure, being a free system allows many programmers to produce viruses for various attacks. Faced with this problem makes it vulnerable. "In the case of Android, the kernel was designed based on version 2.6 of the Linux kernel having similarity in its functionalities, such as security, memory management, process management, etc." (Silva, cited by Leite e Reis [8]). Android is an operating system that has become very popular due to the number of phones sold, causing greater vulnerability regarding its handling if there is no information security culture and carelessness has taken over many users, in addition to other technological security procedures that should be strengthened in the development of iOS systems. According to Munhoz, quoted by Quissanga [1], "A new virus for Android has now emerged and can be doom for many people, because the malicious 'software' automatically performs 'downloads', including of paid 'apps' and games." However, it is possible to have an overview of the risks that there are mobile operating systems, so it is more talked about viruses on computers, in this case, we see the relevance of its study by creating security policies to avoid virtual plagues.

4.2 iOS

Apple's operating system is extremely of Android which is a free system, iOS is closed and has many restrictions to prevent certain pests from affecting the system, manufactured by Apple's kernel, the first iPhone was released in June 2007, and many phones were sold during its launch in the US, being the second in the world with the largest number of devices, has many peculiar features that attract its customers, its graphics and image resolution, the quality of photos are one of the preferences, but we do not mean that only these details, has others for the choice of one or the other, as well as some complaints from some customers, the form of protection of the shop, and the impediment of sending data via "bluetooth," which on the one hand for some disadvantages, others see as being protection, since many viruses contaminate through the Bluetooth connection. However, we cannot say that the system is so robust from the security point of view as can be seen in Pandya and Stamp's quote [9]:

> *It is clear that the iPhone is a vulnerable device with several security flaws. The iPhone's security philosophy itself has a significant flaw. Apple's approach to making the iPhone a secure device was to reduce "the intensity of the device attack 'or' the exposure of the device to vulnerabilities." To do this, Apple allowed write access only to a sandbox area on the file system and impermissible installation of third-party applications.*

The iOS systems allow cyber attacks, although different from Android, but the attacks affect and change the password of the system root, making Crack replicate information to the other host (host) without the user realizing, another way is when the attack is done in the App store, the fastest virus that can destroy the boot system is the Trojan horse very fast in the way of contamination. But the worrying thing is the spies (spywares) that monitor every process of the mobile phone to steal confidential information.

4.3 Describing vulnerabilities in operating systems: Android and Apple iOS

Android is a secure system from the point of view of computer viruses, but it also has some vulnerabilities, however it is important to describe some incorrect practices of attacks and configuration changes:

1. Android is based on the technology and programming languages java, kotlin, C, and C++, there are many developers, which makes it vulnerable;

2. Unlock the Android root;

3. Instal an application in APK format;

4. Android is a system developed by Linux open platform that allows many developers;

5. Bluetooth is one of the fastest ways of transmitting computer viruses;

6. The play store on the mobile phone allows you to activate and instal systems outside of it, although that is not the responsibility of the developer, but of the user.

For Apple's iOS is also a secure operating system, but it is important to mention some incorrect practices:

1. Allow a gaolbreak, this option will allow you to instal applications of unknown origin, which may be malware or spyware;

2. Enable installation of the Unflod Baby Panda malware, which affects jailbroken devices.

However, this seems to be simple information about vulnerability, but it is very important to know the security of the information of both, because they are the most popular phones, and if we compare the damage is fatal, it reminds us of the Panama Papers.

According to the Terra website [10]:

> *The leak of 11.5 million documents—the so-called Panama Papers from Panamanian law and consulting firm Mossack Fonseca, the world's fourth largest offshore law firm, is said to have revealed details of hundreds of thousands of clients using overseas tax havens allegedly for tax evasion, money laundering, drug, and arms trafficking.*

Imagine thefts by espionage by videos, images, information, confidences, or the cloning of the credit card that produce an impact on the average, things that there is no price, in your company as to your reputation, sometimes we can be blackmailed millions of dollars, this does not worry?

4.4 Unsafe practices in mobile phones of systems: android and Apple iOS

In this theme it is important to mention some unsafe practices of users, facilitating the crack to carry out cyber attacks, in this case the customers somehow contribute by not having basic education in information security. In this case the failures are not always related to the company that developed the mobile device, so when we talk about these two systems we can deduce that they are robust and safe regarding the attacks on computer viruses, with some specificities that differ them. However, it is necessary to know how we have failed, since each one of us has used an Android mobile phone or iPhone and knows how it works, but with some limitations of the information security processes, carelessness has been noticed in many users of mobile devices, which translates into huge economic losses, social engineering, cloning of credit cards on unshielded sites, are the focus of Hackers and Crackers to perform computer crime. For Futurelearn [5]: "Cyber security attacks take many forms, from obtaining personal information of users to attacking critical national infrastructure and obtaining proprietary data of companies." However, we can name three forms of attacks:

1. The attack is carried out by identifying the victims online;

2. Generally the attack is on industrial plants;

3. Finally it is done individually by target.

Given this panorama, it is important to conclude this topic by touching on some harmful practices caused by carelessness, lack of knowledge, ignorance, among others, which allow computer theft in mobile phones:

1. Opening suspicious emails (email spoofing, *email spoofing*) with spyware, malware and phishing that captures highly confidential information;

2. Allow bank details to be diverted;

3. Instal software on your phone on sites outside *playstore, appstore, itunesstore and googlestore;*

4. Online purchases on websites of accredited companies, without the possibility of credit card cloning;

5. Make it easier to change data in the phone database without authorization (Pharming);

6. Use a server without *anti-malware, anti-spyware, and firewall*;

7. Facilitating espionage (Sniffing), intersecting and monitoring data for information theft;

8. Allow someone to check while you enter your password;

9. Use social networks without the utmost caution for what you install and configure;

10. Using the mobile device without a secure password.

4.5 Propose security measures in operating systems: android and Apple iOS

"A saying used by people dealing with computer security issues goes as follows: A totally secure system is one that is switched off, locked in a vault and watched over by armed guards"(Ferreira [11]).

Security measures can be in the scope of hardware (Physical) and software (Logical). There are several, but we will mention only a few, and we cannot forget four protocols for information security:

1. Authenticity: the term allows us to accredit that it is authorized. The information must be used by the user and not someone else;

2. Confidentiality: it is exclusively for authentic users;

3. Integrity: the information provided must be passed on as it was designed to its recipient;

4. Intimacy: only authorized persons may have contact with the information on the phone.

4.5.1 Measures can also be preventive, detective, and corrective

1. Preventive measures: these are measures that precede a cyber attack, for example, controlling the entire mobile phone security system, assigning defaults and passwords, creating *back-ups* or redundant backups, installing *anti-spyware* and *anti-spam*; Making online purchases with credit cards on websites of companies that are shielded to avoid cloning: Use secure *websites* and email, make purchases in the appropriate shops: play store, *appstore* and *itunesstore*;

2. Detective measures: this is when we carry out an audit process to analyze the mobile phone's security weakness, the aim is to find possible vulnerabilities or cyberattacks for possible correction;

3. Corrective measures: these are applied when there is any cybersecurity failure, they are solved in an emergent way, to avoid a lot of irreversible damage.

4.5.2 Logical safety policy

1. Assign passwords that are difficult to decipher, with various combinations and changes, to prevent attack (Social Engineering), we suggest password management: KeePass;

2. Use cryptographic techniques: hide codes to present information without being revealed;

3. Recognise the secure web browser showing a padlock and the word S in the link, that is, the non-secure one shows the following: http, the secure one shows: https;

4. Test your web browser using a free Panopticlick tool;

5. We suggest Privacy Badger which is a website blocker in case your website is being targeted by spies;

6. For information protection in case you want to permanently delete unwanted or old information without the possibility of recovering it, we suggest the File Shredder tool;

7. For WiFi technology, modern WPA encryption is recommended.[2]

8. In case of phone calls we propose the Signal app encrypts the calls prevents the communication from being leaked;

9. We suggest you use the VeraCrypt tool to encrypt your important documents;
10. *Firewall*: instal in Noroot firewall, to block suspicious sites or deny, or allow selectively. Very important for information security;
11. *Honey pot*: is a security system tester, also used to preserve the network from cyber attacks;
12. Virtual Private Network (VPN): is using with noroot, are responsible for creating authenticity, privacy and integrity of data, combined with cryptographic technology.

Medranda [12] makes mention about the level of security used in social networks:

> *The first category studied was the level of security of social networks, considering the choice and use of the security password used by students in social networks, 112 students responded that they used the same password for all, but not for email or mobile phone, which represented 67.5% of the total. 21.8% (35 students) they mentioned that it differs depending on the social network and also different from the email address and mobile phone PIN. The lowest percentage corresponded to the use of the password for everything (social networks, email and mobile phone) 8.1%.*

However, we can observe that social networks are now the focus for many contaminations and cyber attacks. According to Tumejormovil [13] "In fact, according to 2017 studies, 67.1% of mobile phones worldwide have Android and, specifically in Spain, 90%, so it is shown that it is a good operating system and could quietly be the best on the market." As we said before by the numbers of users in mobile cellular devices, the control must be jointly, the company does its part in implementing security and the customer must follow the same security protocol to avoid security collapse. For Rina [14] "iOS Mobile Operating System: iOS (formerly called iPhone OS) is from Apple Inc. It has the second largest installed base worldwide in smartphones, but the largest profits due to aggressive price competition between Android-based Manufacturers." Yes, iOS has a lot of audience, despite its restrictions and some bureaucratic processes that on the other hand is a security mechanism that is often not understood, a very simple example the Bluetooth on Android is not possible to send to iOS, the withdrawal of any information on the iPhone is difficult compared to Android.

> *Inside briefly we can explain that these two largest world markets (OS) are evolving and expanding as iOS to include versions of many features that are in Android. Meanwhile, Android gets a technical overhaul that it needs to match iOS in terms of speed and battery life (Haseeb [15]).*

In other words, the Android operating system has many versions and many mobile devices on the market, the two operating systems try to solve some differences, now Android tries to find technology to overcome or match a feature that iOS has the battery life, that there are already also mobile phones Android mobile phones with acceptable battery life.

5. Results and discussion

Android and iOS operating systems regarding computer viruses are safe, but each one has its vulnerability, because we cannot say we are totally safe, from the point of view of comparison with the two operating systems Android and Apple iOS, the latter presents greater protection of the security system, Since Apple's iOS was made by a different programming language than Android, we know that there are specific attacks for a particular operating system, for example, Android tends to be more vulnerable than iOS, by having an open platform that translates into a considered number of developers, otherwise the permission of the installation outside the play store.

Android and iOS operating system are the most preferred of cyber attacks as they have many users.

As for cyber attacks, social networks are now the focus for many contaminations.

In the path of information security in Apple iOS operating system takes a better approach Alvarez [16] "To protect its users from malicious apps, Apple has introduced a vetting process which ensures that all apps follow Apple's privacy rules before they can be made available through the App Store."

In this security theme for iOS, there has been another investment from the PiOS tool, which does statistical analysis to find information from Objective-C source code and Mach-Obinaries data flowcharts. This tool has been successful in checking for information leaks on mobile devices.

We can note that some insecure practices by users make it easier for Crack to carry out cyber attacks. Carelessness has been noted in many users of mobile devices, which causes the cloning of credit cards that translates into economic and financial losses.

It is not always the user who fails, but also some companies do not properly protect customer information.

It is important that companies promote basic training on information security techniques to employees.

However, to have security, it is necessary to use the protocols: authenticity, confidentiality, integrity, and intimacy. As for the measures of always: Preventive, detective, and corrective. Regarding the security policy, we suggest some applications already mentioned previously to be able to protect the computer systems.

6. Conclusions

During the comparative study of mobile cellular operating systems: Android and iOS, it can be concluded that: iOS is more secure compared to Android, due to the restrictions and security mechanisms imposed by Apple. Otherwise, computer virus infections in mobile phones are not very worrying at the moment, but other cyber attacks, spyware that captures confidential information leading to a remote server, that is, even geographically distant, it monitors all user information to take advantage. Meanwhile, we can cite three forms of attacks: the attack is done by identifying the victims online, in industrial, and lastly by targeting them individually. However, the carelessness of users in information security, such as social engineering, attacks on banking data, cloning of credit cards, and the installation of software outside the play store, appstore, and itunesstore are the biggest concern. For this reason, it was necessary to propose information security policies: the greatest care with passwords;

for WiFi technology the modern WPA encryption is recommended; installation in Noroot firewall; use: the Privacy Badger tool to block unsafe sites; the Panopticlick specialist in web testing, VeraCrypt to encrypt important documents and in the case of telephone calls the Signal application that encrypts the calls prevents anyone from listening to the call made.

Author details

Fernando Cassinda Quissanga[1,2]

1 Iberoamerica International University, Mexico

2 European University of the Atlantic, Spain

*Address all correspondence to: fernandoquissanga@hotmail.com

References

[1] Quissanga FC. Characterization of cellular mobile operating systems: Android, Symbian, iphone and Windows phone. Project, Design and Management. 2019;**1**(2):80-83. DOI: 10.35992/pdm.v1i2.200

[2] Creswell JW. Research design qualitative, quantitative and mixed methods. 2010. Available from: https://edisciplinas.usp.br/pluginfile.php/696271/mod_resource/content/1/Creswell.pdf.

[3] Chaer G, Diniz R, Ribeiro E. The questionnaire technique in educational research: The questionnaire in empirical issues. Evidência, Araxá. 2011;7(7):251-266. Available from: http://www.educadores.diaadia.pr.gov.br/arquivos/File/maio2013/sociologi a_artigos/pesqusia_social.pdf

[4] Quissanga FC. Characterization of cellular mobile operating systems: Android, Symbian, iphone and Windows phone. Project, Design and Management. 2019;**1**(2):77. DOI: 10.35992/pdm.v1i2.200

[5] Futurelearn. The National Cyber Security Centre– The Open University; 2019. Available from: https://www.futurelearn.com/courses/introduction-to-cybersecurity/17/steps/565630

[6] Martinelli H. Vírus de celular: Estudo e classificação para um protótipo de defesa: ANEXO B: Código do vírus Cabir. Brasil- Porto Alegre: Uniritter; 2008. Retrieved from: http://www.uniritter.edu.br/graduacao/informatica/sistemas/downloads/tcc2k8/TCC%202%20final(Horst).pdf

[7] Le Thanh H. Analysis of malware families on android mobiles: Detection characteristics recognizable by ordinary phone users and how to fix it. Journal of Information Security. 2013;**4**(4):12. Article ID: 36799. DOI: 10.4236/jis.2013.44024

[8] Leite AC, Reis HM. Comparativo entre sistemas operativos móveis - Android x iOS. 2017. p. 7. Available from: https://simtec.fatectq.edu.br/index.php/simtec/article/view/253/236

[9] Pandya VR, Mark S. iPhone security analysis: Security analysis. Journal of Information Security. 2010;**1**:74-87. DOI: 10.4236/jis.2010.12009. Published Online October 2010. Available from: http://www.SciRP.org/journal/jis

[10] Terra. Panama papers. The Interne. 2016. Available from: https://www.terra.com.br/noticias/mundo/panama-papers-geram-denuncias-e-investigacoes-em-todo-o-mundo,814039f797239995dea030884c41f8faakajlviv.html

[11] Ferreira FC. De Sistemas operativos. Esab - Escola Superior Aberta do Brasil Ltda. First edition: 2008. Available from: http://www.esab.edu.br.

[12] Medranda, J. M. A. Enfoque de las redes sociales en estudantes universitários. 2017. p. 192. Available from: https://dialnet.unirioja.es/servlet/articulo?codigo=6102844.

[13] Tumejormovil Comparison of the most used mobile operating systems (Android, iOS, Windows Phone): Operating System. 2019. Available from: https://tumejormovil.com/operative-systems/.

[14] Rina. A comparative analysis of mobile operating systems. 2018;**6**(12):70 E-ISSN: 2347-2693. Available from: https://www.ijcseonline.org/pub_paper/11-IJCSE-05378.pdf

[15] Haseeb A. Comprehensive and technical overview of Android and IOS OS. International Journal of Computer Sciences and Engineering. 2015;**3**(1):49-57 E-ISSN: 2347-2693. Available from: Comprehensive and Technical Overview of Android and IOS OS (ijcseonline.org)

[16] Alvarez I. A study on Apple's iOS operating system. kriativ-tech. Edição Nº 7 – 28 de Agosto de 2018. ISSN Print: 1646-9976 | ISSN Online: 2184-223X | DOI: 10.31112/kriativ-tech-2018-01-20